SAS® System for Regression

Third Edition

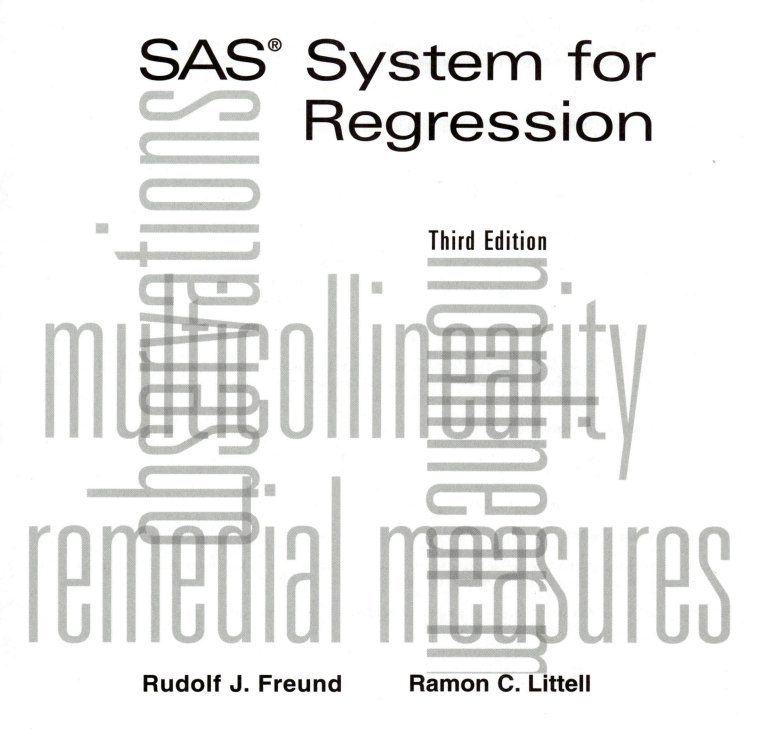

Rudolf J. Freund   **Ramon C. Littell**

Comments or Questions?

The authors assume complete responsibility for the technical accuracy of the content of this book. If you have any questions about the material in this book, please write to the author at this address:

SAS Institute Inc.
Books by Users
Attn: Rudolf J. Freund and Ramon C. Littell
SAS Campus Drive
Cary, NC 27513

If you prefer, you can send e-mail to sasbbu@sas.com with "comments on *SAS System for Regression, Third Edition*" as the subject line, or you can fax the Books by Users program at (919) 677-4444.

The correct bibliographic citation for this manual is as follows: Freund, Rudolf J. and Littell, Ramon C., *SAS® System for Regression, Third Edition,* Cary, NC: SAS Institute Inc., 2000.

SAS® System for Regression, Third Edition

Copyright © 2000 by SAS Institute Inc., Cary, NC, USA.

SAS Institute, Inc. ISBN 1-58025-725-9
John Wiley & Sons, Inc. ISBN

Contents

Acknowledgments

We would like to acknowledge several persons at SAS Institute whose efforts have contributed to the completion of this book. First of all, we are grateful to Jim Goodnight who originally encouraged us to write this book. Several persons reviewed the chapters and contributed many useful comments. The reviewers were Jim Ashton, Jenny Kendall, Charles Lin, Eddie Routten, Warren Sarle, and Mike Stockstill.

The work of several persons has influenced our writing. In particular, we acknowledge Professors Walt Harvey of Ohio State University, Ron Hocking of Texas A&M University, Bill Sanders of the University of Tennessee, and Shayle Searle of Cornell University.

For the Third Edition, we gratefully acknowledge the SAS Institute reviewers Mike Patetta, Sanford Gayle, Lynne Bresler, Brent Cohen, Carole Beam, and Randy Lundberg, who provided many good ideas and pointed out some of our mistakes. A special thanks goes to Eddie Routten who was always available to help clarify many issues. We are also grateful to the outside technical reviewers, Robert L. Mason, Keith E. Muller, and Andrew H. Carp. Also important to this effort is the SAS Institute production team of Maggie Marcum, Mary Rios, and John West. And heartfelt thanks to acquisitions editor Judy Whatley who seemed always to be at her computer waiting to answer our questions.

Finally, we thank the students at Texas A&M University and the University of Florida whose research projects provided the ideas and data for many of the examples

Chapter 1 Regression Concepts

1.1 Statistical Background

Multiple linear regression is a means to express the idea that a response variable, y, varies with a set of independent variables $x_1, x_2, \ldots, x_m$. The variability that y exhibits has two components: a systematic part and a random part. The systematic variation of y can be modeled as a function of the x variables. The model relating y to $x_1, x_2, \ldots, x_m$ is called the *regression equation*. The random part takes into account the fact that the model does not exactly describe the behavior of the response.

Formally, multiple linear regression fits a response variable y to a function of regressor variables and parameters. The general linear regression model has the form

$$y = \beta_0 + \beta_1 x_1 + \ldots + \beta_m x_m + \varepsilon$$

where

  y is the response, or dependent, variable

  $\beta_0, \beta_1, \ldots, \beta_m$ are unknown parameters

  $x_1, x_2, \ldots, x_m$ are the regressor, or independent, variables

  ε is a random error term.

Least squares is a technique used to estimate the parameters based on a set of observed values of these variables. The goal is to find estimates of the parameters $\beta_0, \beta_1, \ldots, \beta_m$ that minimize the sum of the squared differences between the actual y values and the values of y predicted by the equation. These estimates are called the *least-squares estimates,* and the quantity minimized is called the *error sum of squares*.

Typically, you use regression analysis to do the following:

❑   obtain the least-squares estimates of the parameters

❑   estimate the variance of the error term

❑   estimate the standard error of the parameter estimates

❑   test hypotheses about the parameters

❑   calculate predicted values using the estimated equation

❑   evaluate the fit or lack of fit of the model.

The classical linear model assumes that the responses, *y*, are sampled from several populations. These populations are determined by the corresponding values of $x_1, x_2, \ldots, x_m$. As the investigator, you select the values of the *x*'s; they are not random. However, the response values are random. You select the values of the *x*'s to meet your experimental needs, carry out the experiment with the set values of the *x*'s, and measure the responses. Often, though, you cannot control the actual values of the independent variables. In these cases, you should at least be able to assume that they are fixed with respect to the response variable.

In addition, you must assume that

1.   the form of the model is correct; that is, all important independent variables are included and the functional form is appropriate

2.   the expected values of the errors are zero

3.   the variances of the errors (and thus the response variable) are constant across observations

4.   the errors are uncorrelated

5.   for hypothesis testing, the errors are normally distributed.

Not all regression models are necessarily linear in the parameters. For example, the model

$$y = \beta_1 e^{\beta_2 x} + \varepsilon$$

is not linear in the parameter β_2. Specifically, the term $e^{\beta_2 x}$ is not a linear function of β_2. This particular nonlinear model, called the exponential growth or decay model, is used to represent increase (growth) or decrease (decay) over time (*t*) of many types of responses such as population size or radiation counts. Chapter 7, "Nonlinear Models," is devoted to analyses appropriate for this type of model.

Additionally, the random error may not be normally distributed. If this is the case, the least squares technique is not necessarily the appropriate method for estimating the parameters. One such model, the logistic regression model, is presented in Section 7.5.

1.1.1 Terminology and Notation

The principle of least squares is applied to a set of *n* observed values of *y* and the associated x_j to obtain estimates $\hat{\beta}_0, \hat{\beta}_1, \ldots, \hat{\beta}_m$ of the respective parameters $\beta_0, \beta_1, \ldots, \beta_m$. These estimates are then used to construct the fitted model, or estimating equation,

$$\hat{y} = \hat{\beta}_0 + \hat{\beta}_1 x_1 + \ldots + \hat{\beta}_m x_m$$

Many regression computations are illustrated conveniently in matrix notation. Let y_i, x_{ij}, and ε_i denote the values of *y*, x_j, and ε, respectively, in the *i*th observation. The **Y** vector, the **X** matrix, and the ε vector can be defined as follows:

$$\mathbf{Y} = \begin{bmatrix} y_1 \\ . \\ . \\ . \\ y_n \end{bmatrix}, \mathbf{X} = \begin{bmatrix} 1 & x_{11} & . & . & . & x_{1m} \\ . & & . & . & . & . \\ . & . & & & & . \\ . & . & & & & . \\ 1 & x_{n1} & . & . & . & x_{nm} \end{bmatrix}, \mathbf{\varepsilon} = \begin{bmatrix} \varepsilon_1 \\ . \\ . \\ . \\ \varepsilon_n \end{bmatrix}$$

Then the model in matrix notation is

$$\mathbf{Y} = \mathbf{X}\mathbf{\beta} + \mathbf{\varepsilon}.$$

where $\mathbf{\beta}' = (\beta_0\,\beta_1\dots\beta_m)$ is the vector of parameters.

The vector of least-squares estimates is $\hat{\mathbf{\beta}}' = (\hat{\beta}_0, \hat{\beta}_1, \dots, \hat{\beta}_m)$ and is obtained by solving the set of normal equations (NE)

$$\mathbf{X}'\mathbf{X}\mathbf{\beta} = \mathbf{X}'\mathbf{Y}.$$

Assuming that $\mathbf{X}'\mathbf{X}$ is of full rank (nonsingular), the unique solution to the normal equations is given by

$$\hat{\mathbf{\beta}} = (\mathbf{X}'\mathbf{X})^{-1}\mathbf{X}'\mathbf{Y}.$$

The matrix $= (\mathbf{X}'\mathbf{X})^{-1}$ is very useful in regression analysis and is often denoted as follows:

$$(\mathbf{X}'\mathbf{X})^{-1} = \mathbf{C} = \begin{bmatrix} c_{00} & c_{01} & . & . & . & c_{0m} \\ c_{10} & c_{11} & . & . & . & c_{1m} \\ . & . & . & . & . & . \\ . & . & . & . & . & . \\ . & . & . & . & . & . \\ c_{m1} & c_{m2} & . & . & . & c_{mm} \end{bmatrix}$$

1.1.2 Partitioning the Sums of Squares

A basic identity results from least squares, specifically,

$$\Sigma(y - \bar{y})^2 = \Sigma(\bar{y} - \hat{y})^2 + \Sigma(y - \hat{y})^2.$$

This identity shows that the total sum of squared deviations from the mean, $\Sigma(y - \bar{y})^2$, is equal to the sum of squared differences between the mean and the predicted values, $\Sigma(\bar{y} - \hat{y})^2$, plus the sum of squared deviations from the observed y's to the predicted values, $\Sigma(y - \hat{y})^2$. These two parts are called the sum of squares due to regression (or model) and the residual (or error) sum of squares. Thus,

Corrected Total SS = Model SS + Residual SS.

Corrected Total SS always has the same value for a given set of data, regardless of the model that is used; however, partitioning into Model SS and Residual SS depends on the model. Generally, the addition of a new x variable to a model increases the Model SS and, correspondingly, reduces the Residual SS. The residual, or error, sum of squares is computed as follows:

$$\begin{aligned} \text{Residual} \quad \text{SS} &= \mathbf{Y}'\left(\mathbf{I} - \mathbf{X}\left(\mathbf{X}'\mathbf{X}\right)^{-1}\mathbf{X}'\right)\mathbf{Y} \\ &= \mathbf{Y}'\mathbf{Y} - \mathbf{Y}'\mathbf{X}\left(\mathbf{X}'\mathbf{X}\right)^{-1}\mathbf{X}'\mathbf{Y} \\ &= \mathbf{Y}'\mathbf{Y} - \hat{\boldsymbol{\beta}}'\mathbf{X}'\mathbf{Y} \quad . \end{aligned}$$

The error, or residual, mean square

$$s^2 = \text{MSE} = (\text{Residual SS}) / (n - m - 1)$$

is an unbiased estimate of σ^2, the variance of the ε's . This is the so-called error variance generally used in hypothesis testing.

Sums of squares, including the different sums of squares computed by any regression procedure such as the REG and GLM procedures, can be expressed conceptually as the difference between the regression sums of squares for two models, called complete (unrestricted) and restricted models, respectively. This approach relates a given SS to the comparison of two regression models.

For example, denote as SS_1 the regression sum of squares for a complete model with $m=5$ variables:

$$y = \beta_0 + \beta_1 x_1 + \beta_2 x_2 + \beta_3 x_3 + \beta_4 x_4 + \beta_5 x_5 + \varepsilon \quad .$$

Denote as SS_2 the regression sum of squares for a restricted model not containing x_4 and x_5:

$$y = \beta_0 + \beta_1 x_1 + \beta_2 x_2 + \beta_3 x_3 + \varepsilon \quad .$$

Reduction notation can be used to represent the difference between regression sums of squares for the two models:

$$R(\beta_4, \beta_5 \mid \beta_0, \beta_1, \beta_2, \beta_3) = \text{Model SS}_1 - \text{Model SS}_2 \quad .$$

The difference or reduction in error $R(\beta_4, \beta_5 \mid \beta_0, \beta_1, \beta_2, \beta_3)$ indicates the increase in regression sums of squares due to the addition of β_4 and β_5 to the restricted model. It follows that

$$R(\beta_4, \beta_5 \mid \beta_0, \beta_1, \beta_2, \beta_3) = \text{Residual SS}_2 - \text{Residual SS}_1$$

that is, the decrease in error sum of squares due to the addition of β_4 and β_5 to the restricted model. The expression

$$R(\beta_4, \beta_5 \mid \beta_0, \beta_1, \beta_2, \beta_3)$$

is also commonly referred to in the following ways:

❏ the sums of squares due to β_4 and β_5 (or x_4 and x_5) adjusted for $\beta_0, \beta_1, \beta_2, \beta_3$ (or the intercept and x_1, x_2, x_3)

❏ the sums of squares due to fitting x_4 and x_5 after fitting the intercept and x_1, x_2, x_3

❏ the effects of x_4 and x_5 above and beyond or partialing the effects of the intercept and x_1, x_2, x_3.

1.1.3 Hypothesis Testing

Inferences about model parameters are highly dependent on the other parameters in the model under consideration. Therefore, in hypothesis testing, it is important to emphasize the parameters for which inferences have been adjusted. For example, $R(\beta_3 \mid \beta_0, \beta_1, \beta_2)$ and $R(\beta_3 \mid \beta_0, \beta_1)$ may measure entirely different concepts. In other words, a test of $H_0: \beta_3 = 0$ may have one result for the model

$$y = \beta_0 + \beta_1 x_1 + \beta_3 x_3 + \varepsilon$$

and another for the model

$$y = \beta_0 + \beta_1 x_1 + \beta_2 x_2 + \beta_3 x_3 + \varepsilon \quad .$$

Differences reflect actual dependencies among variables in the model rather than inconsistencies in statistical methodology.

Statistical inferences can also be made in terms of linear functions of the parameters of the form

$$H_0: \mathbf{L}\boldsymbol{\beta}_0: \ell_0 \beta_0 + \ell_1 \beta_1 + \ldots + \ell_m \beta_m = 0$$

where the ℓ_i are arbitrary constants chosen to correspond to a specified hypothesis. Such functions are estimated by the corresponding linear function

$$\mathbf{L}\hat{\boldsymbol{\beta}} = \ell_0 \hat{\beta}_0 + \ell_1 \hat{\beta}_1 + \ldots + \ell_m \hat{\beta}_m$$

of the least-squares estimates $\hat{\boldsymbol{\beta}}$. The variance of $\mathbf{L}\hat{\boldsymbol{\beta}}$ is

$$V\left(\mathbf{L}\hat{\boldsymbol{\beta}}\right) = \left(\mathbf{L}(\mathbf{X}'\mathbf{X})^{-1}\mathbf{L}'\right)\sigma^2 \quad .$$

A t test or F test is used to test $H_0: (\mathbf{L}\boldsymbol{\beta}) = 0$. The denominator is usually the residual mean square (MSE). Because the variance of the estimated function is based on statistics computed for the entire model, the test of the hypothesis is made in the presence of all model parameters. Confidence intervals can be constructed to correspond to these tests, which can be generalized to simultaneous tests of several linear functions.

Simultaneous inference about a set of linear functions $\mathbf{L}_1\boldsymbol{\beta}, \ldots, \mathbf{L}_k\boldsymbol{\beta}$ is performed in a related manner. For notational convenience, let $\mathbf{L}$ denote the matrix whose rows are $\mathbf{L}_1, \ldots, \mathbf{L}_k$:

$$\mathbf{L} = \begin{bmatrix} \mathbf{L}_1 \\ \cdot \\ \cdot \\ \cdot \\ \mathbf{L}_k \end{bmatrix}$$

Then the sum of squares

$$\mathrm{SS}(\mathbf{L}\boldsymbol{\beta} = 0) = \left(\mathbf{L}\hat{\boldsymbol{\beta}}\right)' \left(\mathbf{L}(\mathbf{X}'\mathbf{X})^{-1}\mathbf{L}'\right)^{-1}\left(\mathbf{L}\hat{\boldsymbol{\beta}}\right)$$

is associated with the null hypothesis

$$\mathrm{H}_0 : \mathbf{L}_1\boldsymbol{\beta} = \ldots = \mathbf{L}_k\boldsymbol{\beta} = 0 \quad .$$

A test of H_0 is provided by the F statistic

$$F = \left(\mathrm{SS}(\mathbf{L}\boldsymbol{\beta} = 0) / k\right) / \mathrm{MSE} \quad .$$

Three common types of statistical inferences are

❑   a test that all parameters $(\beta_1, \beta_2, \ldots, \beta_m)$ are zero. The test compares the fit of the complete model to that using only the mean:

$$F = (\text{Model SS} / m) / \mathrm{MSE}$$

where

$$\text{Model SS} = \mathrm{R}(\beta_1, \beta_2, \ldots, \beta_m \mid \beta_0)$$

The F statistic has $(m, n-m-1)$ degrees of freedom.[1]

❑   a test that the parameters in a subset are zero. The problem is to compare the fit of the complete model

$$y = \beta_0 + \beta_1 x_1 + \ldots + \beta_g x_g + \beta_{g+1} x_{g+1} + \ldots + \beta_m x_m + \varepsilon$$

to the fit of the restricted model

$$y = \beta_0 + \beta_1 x_1 + \ldots + \beta_g x_g + \varepsilon \quad .$$

---

[1] $\mathrm{R}(\beta_0, \beta_1, \ldots, \beta_m)$ is rarely used. For more information, see the NOINT option in Section 2.4.5.

An F statistic is used to perform the test

$$F = (\mathrm{R}(\beta_{g+1},...,\beta_m \,|\, \beta_0, \beta_1,...,\beta_g) \,/\, (m-g)) \; \mathrm{MSE} \quad .$$

Note that an arbitrary reordering of variables produces a test for any desired subset of parameters. If the subset contains only one parameter, β_m, the test is

$$\begin{aligned} F &= \left(\mathrm{R}\left(\beta_m \,|\, \beta_0, \beta_1, \ldots, \beta_{m-1}\right) / 1\right) / \mathrm{MSE} \\ &= \left(\text{partial SS due to } \beta_m\right) / \mathrm{MSE} \end{aligned}$$

which is equivalent to the t test

$$t = \hat{\beta}_m \,/\, s_{\hat{\beta}_m} = \hat{\beta}_m \,/\, \sqrt{c_{mm}\mathrm{MSE}} \quad .$$

The corresponding $(1 - \alpha)$ confidence interval about β_m is

$$\hat{\beta}_m \pm t_{\alpha/2} \, \sqrt{c_{mm}\mathrm{MSE}} \quad .$$

❏ estimation of a subpopulation mean corresponding to a specific x. For a given set of x values described by a vector $\mathbf{x}$, denote the population mean by $\mu_{\mathbf{x}}$. The point estimate of that population mean is

$$\hat{\mu}_{\mathbf{x}} = \hat{\beta}_0 + \hat{\beta}_1 \mathbf{x}_1 + ... + \hat{\beta}_m \mathbf{x}_m = \mathbf{x}\hat{\boldsymbol{\beta}} \quad .$$

The vector $\mathbf{x}$ is constant; hence, the variance of the estimate, $\hat{\mu}_{\mathbf{x}}$, is

$$\mathrm{V}(\hat{\mu}_{\mathbf{x}}) = \mathbf{x}'(\mathbf{X}'\mathbf{X})^{-1}\mathbf{x}\sigma^2 \; .$$

This equation is useful for computing confidence intervals. A related inference concerns a future single value of y corresponding to a specified x whose estimate is denoted by x. The point estimate is the same as that for the mean, but its variance is

$$\mathrm{V}(\hat{y}_{\mathbf{x}}) = (1 + \mathbf{x}'(\mathbf{X}'\mathbf{X})^{-1}\mathbf{x})\sigma^2 \; .$$

1.1.4 Using the Generalized Inverse

Many applications of regression procedures involve an $\mathbf{X}'\mathbf{X}$ matrix that is not of full rank and has no unique inverse. PROC GLM and PROC REG compute a generalized inverse $(\mathbf{X}'\mathbf{X})^-$ and use it to compute a regression estimate

$$\mathbf{b} = (\mathbf{X}'\mathbf{X})^- \mathbf{X}'\mathbf{Y} \quad .$$

A generalized inverse of a matrix $\mathbf{A}$ is any matrix $\mathbf{G}$ such that $\mathbf{AGA=A}$. Note that this also identifies the inverse of a full-rank matrix.

If $\mathbf{X'X}$ is not of full rank, then an infinite number of generalized inverses exist. Different generalized inverses lead to different solutions to the normal equations that have different expected values; that is, $E(\mathbf{b}) = (\mathbf{X'X})^{-}\mathbf{X'Y}\boldsymbol{\beta}$ depends on the particular generalized inverse used to obtain $\mathbf{b}$. Therefore, it is important to understand what is being estimated by the solution.

Fortunately, not all computations in regression analysis depend on the particular solution obtained. For example, the error sum of squares is invariant with respect to $(\mathbf{X'X})^{-}$ and is given by

$$\text{SSE} = \mathbf{Y'}\!\left(\mathbf{1} - \mathbf{X}(\mathbf{X'X})^{-}\mathbf{X'}\right)\!\mathbf{Y} \quad .$$

Hence, the model sum of squares also does not depend on the particular generalized inverse obtained.

The generalized inverse has played a major role in the presentation of the theory of linear statistical models, notably in the work of Graybill (1976) and Searle (1971). In a theoretical setting it is often possible, and even desirable, to avoid specifying a particular generalized inverse. To apply the generalized inverse to statistical data using computer programs, a generalized inverse must actually be calculated. Therefore, it is necessary to declare the specific generalized inverse being computed. For example, consider an $\mathbf{X'X}$ matrix of rank k that can be partitioned as

$$\mathbf{X'X} = \begin{bmatrix} \mathbf{A}_{11} & \mathbf{A}_{12} \\ \mathbf{A}_{21} & \mathbf{A}_{22} \end{bmatrix}$$

where $\mathbf{A}_{11}$ is $k \times k$ and of rank k. Then $\mathbf{A}_{11}^{-1}$ exists, and a generalized inverse of $\mathbf{X'X}$ is

$$(\mathbf{X'X})^{-} = \begin{bmatrix} \mathbf{A}_{11}^{-1} & \varphi_{12} \\ \varphi_{21} & \varphi_{22} \end{bmatrix}$$

where each φ_{ij} is a matrix of zeros of the appropriate dimension.

This approach to obtaining a generalized inverse, the method used by PROC GLM and PROC REG, can be extended indefinitely by partitioning a singular matrix into several sets of matrices as illustrated above. Note that the resulting solution to the normal equations, $\mathbf{b}=(\mathbf{X'X})^{-}\mathbf{X'Y}$, has zeros in the positions corresponding to the rows filled with zeros in $(\mathbf{X'X})^{-}$. This is the solution printed by these procedures, and it is regarded as providing a biased estimate of $\boldsymbol{\beta}$.

However, because $\mathbf{b}$ is not unique, a linear function, $\mathbf{Lb}$, and its variance are generally not unique either. However, a class of linear functions called *estimable functions* exists, and they have the following properties:

❏   The vector $\mathbf{L}$ is a linear combination of rows of $\mathbf{X}$.

❏ $\mathbf{Lb}$ and its variance are invariant through all possible generalized inverses. In other words, $\mathbf{Lb}$ is unique and is an unbiased estimate of $\mathbf{L}\boldsymbol{\beta}$.

Analogous to the full-rank case, the variance of an estimable function $\mathbf{Lb}$ is given by

$$V(\mathbf{Lb}) = (\mathbf{L}\,(\mathbf{X'X})^{-}\mathbf{L'})\sigma^2$$

This expression is used for statistical inference. For example, a test of $H_0 : \mathbf{L\beta} = 0$ is given by the t test

$$t = \mathbf{Lb} / \sqrt{\left(\mathbf{L(X'X)^- L'}\right)}\mathrm{MSE} \ .$$

Simultaneous inferences on a set of estimable functions are performed in an analogous manner.

1.2 Performing a Regression with the IML Procedure

As you can see in Section 1.3, "Regression with the SAS System," and in greater detail in subsequent chapters, the SAS System provides a flexible array of procedures for performing regression analyses. You can also perform these analyses by direct application of the matrix formulas presented in the previous section using SAS/IML software. This software, which is implemented as PROC IML, is most frequently used for the custom programming of methods too specialized or too new to be packaged into the standard regression procedures. It is also useful as an instructional tool for illustrating linear model and other methodologies.

The following example represents a regression analysis performed by PROC IML. This example is not intended to serve as a tutorial in the use of PROC IML. If you need more information on PROC IML, see the *SAS/IML User's Guide*. The example for this section is the one used in Chapter 2, "Using the REG Procedure," to illustrate PROC REG. The data set is described, and the data are presented in Section 2.1, "Introduction." For this presentation, the variable CPM is the dependent variable y, and the variables UTL, SPA, ALF, and ASL are the independent variables x_1, x_2, x_3, and x_4, respectively. Comment statements are used in the SAS program to explain the individual steps in the analysis.

```
/* Invoke PROC IML and create the x and y matrices using  */
/* the variables UTL, SPA, ALF, and CPM from the SAS data  */
/* set AIR.                                                 */

proc iml;
   use air;
   read all var {'utl' 'spa' 'alf' 'asl'} into x;
   read all var {'cpm'} into y;
/* Define the number of observations (N) and the number of  */
/* variables (M) as the number of rows and columns of X.    */
/* Add a column of ones for the intercept variable to the X */
/* matrix.                                                   */

   n=nrow(x);      /* number of observations */
   m=ncol(x);      /* number of variables    */
   x=j(n,1,1)||x;  /* add column of ones to X */

/* Compute C, the inverse of X'X and the vector of   */
/* coefficient estimates BHAT.                        */

   c=inv(x'*x);
   bhat=c*x'*y;
```

```
/* Compute SSE, the residual sum of squares, and MSE, the   */
/* residual mean square (variance estimate).                */

   sse= y'*y-bhat'*x'*y;
   dfe= n-m-1;
   mse=sse/dfe;

/* The test for the model can be restated as a test for     */
/* the linear function L where L is the matrix.             */

   l={0 1 0 0 0,
      0 0 1 0 0,
      0 0 0 1 0,
      0 0 0 0 1};

/* Compute SSMODEL and MSMODEL and the corresponding F      */
/* ratio.                                                   */

   ssmodel=(l*bhat)'*inv(l*c*l')*(l*bhat);
   msmodel=ssmodel/m;
   f=(ssmodel/m)/mse;

/* Concatenate results into one matrix.                     */
   source=(m||ssmodel||msmodel||f)//(dfe||sse||mse||{.});
/* Compute                                                  */
/* SEB    vector of standard errors of the estimated        */
/*        coefficients                                      */
/* T      matrix containing the t statistic for testing that */
/*        each coefficient is zero                          */
/* PROBT  significance level of test                        */
/* STATS  matrix which contains as its columns the          */

/*        coefficient estimates, their standard errors,     */
/*        and the t statistics.                             */

   seb=sqrt(vecdiag(c)#mse);
   t=bhat/seb;
   probt=2*(1 - cdf('t',abs(t),dfe));
   stats=bhat||seb||t||probt;

/* Compute                                                  */
/* YHAT   predicted values                                  */
/* RESID  residual values                                   */
/* OBS    matrix containing as its columns the actual,      */
/*        predicted, and residual values, respectively.     */

   yhat=x*bhat;
   resid=y-yhat;
   obs=y||yhat||resid;

/* Print the matrices containing the desired results.       */
   print 'Regression Results',
   source (|colname={DF SS MS F} rowname={MODEL ERROR}
   format=8.4|),,
   'Parameter Estimates',
   stats (|colname={BHAT SEB T PROBT} rowname={INT UTL SPA ALF
   ASL}
   format=8.4|) ,,,
   'RESIDUALS', obs (| colname={Y YHAT RESID} format=8.3|) ;
```

The results of this sample program are shown in Output 1.1.

```
                            Regression Results
                            SOURCE
                    DF        SS        MS         F

           MODEL   4.0000   6.5712    1.6428   10.5560
           ERROR  28.0000   4.3575    0.1556      .

                        Parameter Estimates
                            STATS
                   BHAT      SEB         T      PROBT

           INT    8.5955   0.9028    9.5212    0.0000
           UTL   -0.2128   0.0651   -3.2697    0.0029
           SPA   -4.9503   1.2170   -4.0678    0.0004
           ALF   -7.2114   1.3206   -5.4608    0.0000
           ASL    0.3328   0.1813    1.8351    0.0771

                        RESIDUALS
                  OBS   Y     YHAT      RESID

                  2.258    2.574    -0.316
                  2.275    2.136     0.139
                  2.341    3.440    -1.099
                  2.357    2.424    -0.067
                  2.363    2.563    -0.200
                  2.404    2.879    -0.475
                  2.425    2.290     0.135
                  2.711    2.765    -0.054
                  2.743    3.367    -0.624
                  2.780    2.873    -0.093
                  2.833    2.636     0.197
                  2.846    3.183    -0.337
                  2.906    3.190    -0.284
                  2.954    2.932     0.022
                  2.962    2.975    -0.013
                  2.971    3.019    -0.048
                  3.044    3.324    -0.280
                  3.096    2.752     0.344
                  3.140    3.094     0.046
                  3.306    3.569    -0.263
                  3.306    2.748     0.558
                  3.311    3.483    -0.172
                  3.313    3.237     0.076
                  3.392    3.443    -0.051
                  3.437    3.520    -0.083
                  3.462    3.245     0.217
                  3.527    3.149     0.378
                  3.689    3.644     0.045
                  3.760    3.488     0.272
                  3.856    3.565     0.291
                  3.959    3.520     0.439
                  4.024    3.158     0.866
                  4.737    4.302     0.435
```

When you use PROC IML, all results are in the form of matrices. Each matrix is identified by its name, and its elements are identified by row and column indices. You may find it necessary to refer to the program to identify specific elements.

The results of this analysis are discussed thoroughly in Chapter 2; therefore, in this section only the results that can be compared with those from PROC REG (shown in Output 2.5) are identified.

In Output 1.1, the first matrix corresponds to overall model statistics produced by PROC REG. Included here are the degrees of freedom, sums of squares, and mean square for the model and for the error. The *F* statistic tests the significance of the entire model, which includes the independent variables UTL, SPA, ALF, and ASL.

The matrix **STATS** contains the information on the parameter estimates. Rows correspond to parameters (intercept and independent variables UTL, SPA, ALF, ASL, respectively), and columns correspond to the different statistics. The first column contains the coefficient estimates (from matrix **BHAT**), the second contains the standard errors of the estimates (from matrix **SEB**), and the third contains the *t* statistics (from matrix **T**). The final column (from matrix **PROBT**) contains the probability associated with the *t* statistic.

The matrix **OBS** contains the information on observations. The rows correspond to the observations. The first column contains the original *y* values (matrix **Y**), the second contains the predicted values (from matrix **YHAT**), and the third contains the residuals (from matrix **RESID**).

The results achieved by using PROC IML agree with those from PROC REG, as shown in Output 2.5. Because PROC IML is most frequently used for the custom programming of new or specialized methods, the standard regression procedures are more efficient with respect to both programming time and computing time. For this reason, you should try to use these procedures whenever possible. In addition, the output produced with the standard regression procedures is designed to present analysis results more clearly than the printed matrices produced with PROC IML. See Section 1.3 for an overview of standard regression procedures.

1.3 Regression with the SAS System

This section reviews the following SAS/STAT software procedures that are used for regression analysis:

CALIS	ORTHOREG
CATMOD	PLS
GENMOD	PROBIT
GLM	REG
LIFEREG	RSREG
LOESS	TPSPLINE
LOGISTIC	TRANSREG
NLIN	

PROC REG provides the most general analysis capabilities; the other procedures give more specialized analyses. This section also briefly mentions several procedures in SAS/ETS software.

Many SAS/STAT procedures, each with special features, perform regression analysis. The following procedures perform at least one type of regression analysis:

CALIS   fits systems of linear structural equations with latent variables and path analysis.

CATMOD   analyzes data that can be represented by a contingency table. PROC CATMOD fits linear models to functions of response frequencies and can be used for loglinear models and logistic regression.

GENMOD   fits generalized linear models. PROC GENMOD is especially suited for responses with discrete outcomes, and it performs logistic regression and Poisson regression as well as fitting generalized estimating equations for repeated measures data.

GLM   uses the method of least squares to fit general linear models. In addition to many other analyses, PROC GLM can perform simple, multiple, polynomial, and weighted regression, as well as analysis of variance and analysis of covariance. PROC GLM has many of the same input/output capabilities as PROC REG but does not provide as many diagnostic tools or allow interactive changes in the model or data.

LIFEREG   fits parametric models to failure-time data that may be right-, left-, or interval-censored. These types of models are commonly used in survival analysis.

LOESS   fits a response curve or plane to data without using a specified model.

LOGISTIC   fits logistic regression models. PROC LOGISTIC can perform stepwise regressions as well as compute regression diagnostics.

NLIN   fits nonlinear regression models. Several different iterative methods are available.

ORTHOREG   performs regression using the Gentleman-Givens computational method. For ill-conditioned data, PROC ORTHOREG can produce more accurate parameter estimates than other procedures such as PROC GLM and PROC REG.

PLS   performs partial least squares regression, principal components regression, and restricted rank regression, with cross validation for the number of components.

PROBIT   performs probit regression as well as logistic regression and ordinal logistic regression. PROC PROBIT is useful when the dependent variable is either dichotomous or polychotomous and the independent variables are continuous.

REG   performs linear regression with many diagnostic capabilities, selects models using one of nine selection methods, produces scatter plots of raw data and statistics, highlights scatter plots to identify particular observations, and allows interactive changes in both the regression model and the data used to fit the model.

                PROC REG provides options for special estimates, outlier and specification error detection (row diagnostics), collinearity statistics (column diagnostics), and tests of linear functions of parameter estimates. It can perform restricted least-squares estimation and multivariate tests. It can also produce SAS data sets containing the parameter estimates and most of the statistics produced by the procedure.

RSREG   builds quadratic response-surface regression models to determine the factor levels of optimum response, and it performs a ridge analysis to search for the region of optimum response.

TPSPLINE   fits a response curve or plane to data without using a specified model.

TRANSREG   obtains optimal linear and nonlinear transformations of variables using alternating least squares. PROC TRANSREG creates an output data set containing the transformed variables.

SAS/ETS software provides tools for economic analysis and modeling, time-series analysis, and forecasting. Since many of these tools are forms of regression, many procedures in this software also perform regression. These include the following:

AUTOREG   implements regression models using time-series data where the errors are autocorrelated.

MODEL   handles nonlinear simultaneous systems of equations, such as econometric models.

PDLREG   performs regression analysis with polynomial distributed lags.

SYSLIN   handles linear simultaneous systems of equations, such as econometric models.

TSCSREG   handles regression models that use both time-series and cross-sectional data.

Finally, if a regression method cannot be performed by any of the SAS procedures above, SAS/IML software provides an interactive matrix language than can be used, as shown in Section 1.2.

Chapter 2 Using the REG Procedure

2.1 Introduction

As indicated in Section 1.3, "Regression with the SAS System," PROC REG is the primary SAS procedure that performs the computations for a statistical analysis of data based on a linear regression model. The basic statements for performing such an analysis are

```
proc reg;
    model list of dependent variables = list of independent
          variables
    /model options;
```

This chapter provides instructions on the use of PROC REG for performing regression analyses. Included are instructions for employing some of the more frequently used options, providing for informative plots, producing data sets for further analysis, and utilizing the interactive features available for this procedure. Subsequent chapters deal with procedures for more specialized analyses and models.

The data for the example in this chapter concern factors considered to be influential in determining the cost of providing air service. The goal is to develop a model for estimating the cost per passenger mile so that the major factors in determining that cost can be isolated. The source of the data is a Civil Aeronautics Board report, *Aircraft Operating Cost and Performance Report* (August 1972). The variables are

CPM   cost per passenger mile (cents)

UTL   average hours per day use of aircraft

ASL   average length of nonstop legs of flights (1000 miles)

SPA   average number of seats per aircraft (100 seats)

ALF   average load factor (% of seats occupied by passengers).

Data have been collected for 33 U.S. airlines with average nonstop lengths of flights greater than 800 miles.

An additional indicator variable, TYPE, has been constructed. This variable has value zero for airlines with ASL<1200 miles, and unity for airlines with ASL$\geq$1200 miles. This variable may be constructed in the DATA step with a set of IF-THEN statements. Alternately, you can use the Boolean operator as follows:

```
type=(asl>=1.200);
```

which assigns the value one to the variable TYPE if the condition in the parentheses is true and zero otherwise. This variable is used in Section 2.10, "Predicting to a Different Set of Data." The data set, named AIR, appears in Output 2.1.

Output 2.1
Airline Cost
Data

Obs	ALF	UTL	ASL	SPA	TYPE	CPM
1	0.591	7.87	1.790	0.1375	1	2.258
2	0.488	9.50	2.515	0.3546	1	2.275
3	0.412	7.91	1.350	0.1920	1	2.341
4	0.397	13.30	3.607	0.3390	1	2.357
5	0.582	8.48	1.963	0.1381	1	2.363
6	0.466	9.38	1.123	0.1481	0	2.404
7	0.535	10.80	1.576	0.1361	1	2.425
8	0.434	8.36	1.912	0.3148	1	2.711
9	0.439	8.43	1.584	0.1607	1	2.743
10	0.417	8.83	2.377	0.3287	1	2.780
11	0.400	8.42	1.495	0.3597	1	2.833
12	0.410	9.62	0.840	0.1390	0	2.846
13	0.478	8.71	1.392	0.1148	1	2.906
14	0.495	8.44	0.871	0.1186	0	2.954
15	0.476	8.91	0.961	0.1236	0	2.962
16	0.539	6.84	1.008	0.1150	0	2.971
17	0.409	9.00	0.845	0.1390	0	3.044
18	0.381	10.20	1.692	0.3007	1	3.096
19	0.486	8.29	0.877	0.1060	0	3.140
20	0.287	8.09	1.528	0.3522	1	3.306
21	0.504	9.47	1.408	0.1345	1	3.306
22	0.455	7.70	1.236	0.1221	1	3.311
23	0.405	9.57	0.863	0.1390	0	3.313
24	0.422	8.35	1.031	0.1365	0	3.392
25	0.476	7.27	1.416	0.1145	1	3.437
26	0.426	7.52	0.975	0.2025	0	3.462
27	0.349	9.56	2.189	0.3279	1	3.527
28	0.394	7.94	0.949	0.1488	0	3.689
29	0.452	7.55	1.164	0.1270	0	3.760
30	0.425	10.60	2.780	0.1282	1	3.856
31	0.362	10.80	1.518	0.1356	1	3.959
32	0.541	6.31	0.823	0.0943	0	4.024
33	0.378	5.65	0.821	0.1290	0	4.737

2.2 A Model with One Independent Variable

Regression with a single independent variable is known as a simple linear regression model. Since such a model is easy to visualize, it is presented here to introduce several aspects of regression analysis that you can perform with the SAS System. Using the single variable ALF, the average load factor, to estimate the cost per passenger mile, CPM, the model is

$$CPM = \beta_0 + \beta_1(ALF) + \varepsilon \quad .$$

In this model, β_1 is the effect on the per passenger cost of a one unit (percentage point) increase in the load factor. For this example, you expect this coefficient to be negative. The coefficient β_0, the intercept, is the cost per passenger mile if the load factor is zero. Since a zero load factor is not possible, the value of this coefficient is not useful, but the term is needed to fully specify the regression line. The term ε represents the random error and accounts for variation in cost due to factors other than variation in ALF.

In the case of regression with one independent variable, it is instructive to plot the observed variables. The SAS System provides plotting capabilities with either PROC PLOT, which produces line-printer plots, or PROC GPLOT, a SAS/GRAPH procedure that produces high-resolution plots. Both of these will be used in this book; you can use either for virtually all examples. You can produce a line printer plot with the following SAS statements:

```
proc plot data=air;
   plot cpm*alf / hpos=30 vpos=25;
run;
```

The HPOS= and VPOS= options specify the size of the plot; the default is a full-page plot. Output 2.2 shows the result.

Output 2.2
Plot of Cost against Load Factor

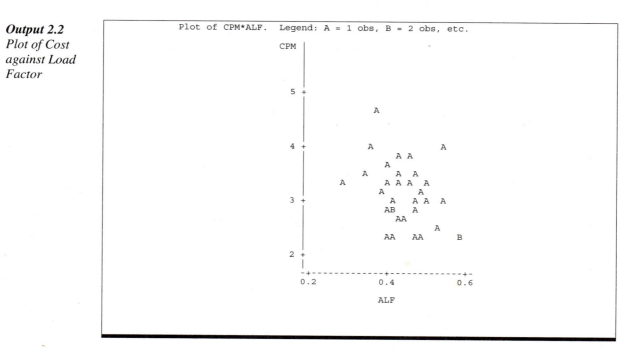

The plot shows the expected tendency for lower costs (CPM) with higher load factors (ALF). However, the relationship is not very strong. This suggests that there may be other factors affecting CPM as you will see in Section 2.3, "A Model with Several Independent Variables."

The SAS statements required for performing a regression analysis for this model are

```
proc reg data = air;
   model cpm = alf;
run;
```

The PROC statement invokes the REG procedure, and DATA=AIR tells the SAS System to use the data set AIR. (The SAS System applies the called procedure to the most recently created data set if DATA= *data set name* is not specified in the PROC statement, but the use of this default is not recommended.)

The MODEL statement contains the equation CPM=ALF, to specify the desired model. The left-hand side specifies the dependent variable (CPM), and the right-hand side specifies the independent variable (in this case ALF). The intercept term is not specified because PROC REG automatically assumes that it is to be estimated (unless the NOINT option is used; see Section 2.4.5, "Regression through the Origin: the NOINT Option"). Also, no term is specified in the MODEL statement corresponding to the error term ε. PROC REG will produce ordinary least-squares estimates of the parameters (β_0 and β_1 in this example), which are optimal if the errors are independent and have equal variances. Methods for checking these assumptions and some suggestions for alternate methodologies are presented in Chapter 3, "Observations."

You can use LABEL statements and/or, starting with Version 7 of the SAS System, variable names with more than 8 characters to provide more easily interpretable output. Because these are essentially cosmetic options and do not affect methodology or results, they will not be employed in this book.

The output from PROC REG appears in Output 2.3. The circled numbers in Output 2.3 have been added to key the descriptions that follow.

Output 2.3 Results for Regression with One Independent Variable

```
                          The REG Procedure
                    ①    Model: MODEL1
                      Dependent Variable: CPM

                        Analysis of Variance

                                  Sum of        Mean        ⑤
   ②  Source          ② DF    ③ Squares   ④ Square    F Value    Pr > F

      Model               1      1.52682     1.52682     5.03     0.0321
      Error              31      9.40185     0.30329
      Corrected Total    32     10.92867

                                                 ⑦
         ⑥ Root MSE              0.55071    R-Square     0.1397
             Dependent Mean      3.10570    Adj R-Sq     0.1120
             Coeff Var          17.73237

                          Parameter Estimates

                      ⑨ Parameter   ⑩Standard      ⑪
     ⑧ Variable   DF     Estimate      Error     t Value    Pr > |t|

        Intercept   1     4.56055      0.65546     6.96      <.0001
        ALF         1    -3.26355      1.45453    -2.24      0.0321
```

① The first two lines identify the model. Since PROC REG may have many models in one execution of the procedure (see Section 2.5.2, "Modifying the Model or the Data"), by default models are labeled MODEL1, MODEL2, and so forth.[1] Since a model may have several dependent variables, the second line identifies the dependent variable.

② These two columns give the sources of variation and degrees of freedom associated with the sums of squares discussed in number ③ below.

③ The regression sum of squares (called Model SS; see Section 1.1.2, "Partitioning the Sums of Squares") is 1.52682, and the residual sum of squares (called Error SS; see Section 1.1.2) is 9.40185. The sum of these two is the Total SS of 10.92867, the corrected total sum of squares. This illustrates the basic identity in regression analysis that Corrected Total SS = ModelSS + ErrorSS, which says that variation among the observed values of the dependent variable can be attributed to two sources: (1) variation that is due to changes in the independent variable and (2) variation that is not due to changes in the independent variable. If the model is correctly specified, this latter variation is random variation.

④ Mean squares are computed by dividing the sums of squares by their respective degrees of freedom. The MS for error (MSE) is an unbiased estimate of σ^2, the variance of ε, if the model has been correctly specified.

⑤ The value of the F statistic, 5.03, is the ratio of the model mean square divided by the error mean square. For the general multiple regression case, it is used to test the composite hypothesis that all coefficients, not including the intercept, are simultaneously zero. In this case, the hypothesis is $\beta_1 = 0$. The p value (which follows the F statistic) of 0.0321 indicates that there is about a 0.032 chance of obtaining an F value this large or larger if, in fact, $\beta_1 = 0$. Thus, there is reasonably strong evidence to state that $\beta_1 \neq 0$.

⑥ Root MSE = 0.55071 is the square root of the error mean square. It estimates the standard deviation of ε. Dependent Mean = 3.10570 is the mean of the dependent variable CPM.[2] Coeff Var = 17.73237 is the coefficient of variation. This is a measure of relative variation and is the ratio of Root MSE to the Dependent Mean, expressed as a percentage. In this example, the error standard deviation is 17.73% of the overall mean value of CPM. The coefficient is sometimes used as a standard to gauge the relative magnitude of the random error.

⑦ R-Square = 0.1397 is the square of the multiple correlation coefficient. For a one-variable regression it is equivalent to the square of the correlation between the dependent and independent variables. Equivalently, since the predicted values are a linear function of the independent variable, it is also the square of the correlation between the observed and predicted values of the dependent variable. Finally, it is also the ratio of Model SS divided by Total SS, and it thereby represents the fraction of total variation in the values of CPM explained by, or due to, the linear relationship to ALF (see number ③ above). Adj R-Sq is an alternative to R-Square, which is discussed in Section 2.3.

---

[1] You may specify your own model identification by preceding the MODEL statement with a label followed by a colon. This is useful if a number of different models are being analyzed.

[2] All of these descriptive statistics are based on the number of observations used in the analysis, which consists of observations having all values of the variables in the MODEL and/or VAR statements. Thus, it may not be the same as the total number of observations in the data set if there are missing values.

⑧ The Variable column identifies the regression coefficients. The label Intercept identifies β_0, and the other coefficients are identified by their respective variable names. In this case, the only coefficient is that for ALF.

⑨ The values in the Parameter Estimate column are the estimated coefficients. The estimate of the intercept, β_0, is 4.56055, and the estimate of the coefficient, β_1, is −3.26355 which shows that on the average, the estimated average cost is decreased by 3.264 cents for each percentage point increase in the average load factor. These give the fitted model

$$\widehat{\text{CPM}} = 4.56055 - (3.26355)(\text{ALF}),$$

where the caret (^) over CPM indicates that it is an estimated value. This expression can be used to estimate the average passenger mile cost for a given average load factor. The following sections show how to print and plot such estimates and their standard errors and confidence intervals.

⑩ The estimated standard errors of the coefficient estimates are given in the Standard Error column. These are 0.65546 and 1.45453, respectively, and may be used to construct confidence intervals for these model parameters. For example, a 95% confidence interval for β_1 is computed as

$$(-3.26355 \pm 2.042*1.45453)$$

where 2.042 is the 0.05 level-two tail value of the *t* distribution for 30 degrees of freedom (used to approximate 31 degrees of freedom needed here).[3] The resulting interval is from −0.293 to −6.235. Thus, with 95% confidence, you can state that the interval (−0.293 to −6.235) includes the true change in cost due to a one-percentage point increase in ALF.

⑪ The *t* statistics, used for testing the individual null hypotheses that each coefficient is zero, are given in the column labeled t Value. These quantities are simply the parameter estimates divided by their standard errors. The next column, labeled Prob> | t |, gives the *p* value for the two-tailed test. For example, the *p* value of 0.0321 for ALF states that if you reject H$_0$: (ALF=0), there is a 0.0321 chance of erroneous rejection. Note that for this one-variable model, this is the same as that for the *F* test of the model. Because the intercept has no practical value in this example, the *p* value for this parameter is not useful.

2.3 A Model with Several Independent Variables

The same example is used throughout this chapter for illustrating the use of PROC REG in a model with several independent variables. This section presents the results obtained from a model for relating cost per passenger mile to all the variables in the data except for TYPE (see Output 2.1).

As you will see later, examining relationships among individual pairs of variables is not often useful in establishing the basis for a multiple regression model. Nevertheless, you may want to examine pairwise correlations among all variables in the model in order to, for example, compare

---

[3] The exact value can be obtained by the TINV function available as a DATA step function.

the strengths of the individual variables with the response to those obtained by using all variables simultaneously in a regression model. In addition, you may want to have some information about the relationships among the variables in the model. The correlations, as well as some simple statistics, are obtained with the statements:

```
proc corr data=air;
   var alf utl asl spa cpm;
run;
```

The output from these statements appears in Output 2.4.[4]

Output 2.4
Correlation among Variables

```
                              The CORR Procedure

            5  Variables:  ALF      UTL       ASL       SPA      CPM


                              Simple Statistics

Variable       N        Mean      Std Dev        Sum       Minimum      Maximum

ALF           33     0.44579      0.06693     14.71100     0.28700      0.59100
UTL           33     8.71727      1.44903    287.67000     5.65000     13.30000
ASL           33     1.46906      0.64998     48.47900     0.82100      3.60700
SPA           33     0.18358      0.08970      6.05810     0.09430      0.35970
CPM           33     3.10570      0.58440    102.48800     2.25800      4.73700


                 Pearson Correlation Coefficients, N = 33
                     Prob > |r|  under H0: Rho=0

                  ALF           UTL           ASL           SPA           CPM

ALF           1.00000      -0.20538      -0.08719      -0.49490      -0.37378
                             0.2515        0.6295        0.0034        0.0321

UTL          -0.20538       1.00000       0.62842       0.32442      -0.37197
              0.2515                      <.0001         0.0655        0.0330

ASL          -0.08719       0.62842       1.00000       0.60710      -0.35078
              0.6295         <.0001                      0.0002        0.0453

SPA          -0.49490       0.32442       0.60710       1.00000      -0.29758
              0.0034         0.0655        0.0002                      0.0926

CPM          -0.37378      -0.37197      -0.35078      -0.29758       1.00000
              0.0321         0.0330        0.0453        0.0926
```

The upper portion of the output reports the mean, standard deviation, and other statistics for the specified variables. The correlation coefficients appear as a matrix in the lower portion of the output. Each row and column heading corresponds to a variable in the VAR (or VARIABLES) list. For a given row and column there are two numbers: the upper number is the estimated correlation coefficient between the row variable and the column variable, and the lower number is the significance probability (*p* value) for testing the null hypothesis that the corresponding population correlation is zero.

---

[4] PROC CORR is a comprehensive SAS procedure for computing various types of correlation and covariance statistics. Only the simplest options are used here. For more details see the *SAS Procedures Guide*.

An interesting feature of the CORR procedure is that each individual correlation coefficient is computed from all available pairs of values. If there are missing values, the number of such pairs is given for each pair below the *p* value. This feature can be very useful for detecting the existence of missing values.[5]

Examining the correlations, you can, for example, see that the estimated correlation between ALF and SPA is −0.4949, and is significantly different from zero at the *p* = 0.0034 level. In other words, this test provides evidence of a negative relationship between the average size of planes and the load factor.

The correlations between the dependent variable, CPM, and the four independent variables are of particular interest. All of these correlations are negative, and all, except for that with SPA, are significantly different from zero (p<0.05). This suggests that each of these three factors can be useful, by itself, in estimating the cost of providing air service.[6]

In terms of the example, the model for estimating CPM using the four cost factors is

$$CPM = \beta_0 + (\beta_1)(ALF) + (\beta_2)(UTL) + (\beta_3)(ASL) + (\beta_4)(SPA) + \varepsilon \quad .$$

To fit this model, use the following statements:

```
proc reg data=air;
   model cpm = alf utl asl spa;
run;
```

The results appear in Output 2.5.

Output 2.5
PROC REG
with CPM
and Four
Independent
Variables

```
                                    The REG Procedure
                                     Model: MODEL1
                                 Dependent Variable: CPM

                                   Analysis of Variance

                                          Sum of        Mean       ①
            Source               DF       Squares      Square     F Value    Pr > F

            Model                 4       6.57115     1.64279     10.56      <.0001
            Error                28       4.35752     0.15563
            Corrected Total      32      10.92867

                                                         ②
                    Root MSE             0.39449     R-Square     0.6013
                    Dependent Mean       3.10570     Adj R-Sq     0.5443
                    Coeff Var           12.70228

                                   Parameter Estimates
                                      ③          ④          ⑤
                                  Parameter    Standard
            Variable      DF       Estimate      Error    t Value    Pr > |t|
            Intercept      1        8.59553     0.90278      9.52     <.0001
            ALF            1       -7.21137     1.32056     -5.46     <.0001
            UTL            1       -0.21282     0.06509     -3.27      0.0029
            ASL            1        0.33277     0.18133      1.84      0.0771
            SPA            1       -4.95030     1.21695     -4.07      0.0004
```

---

[5]   You can use the NOMISS option in the PROC CORR statement to get correlations that correspond to what PROC REG uses.
[6]   Note that the correlation between CPM and ALF is the square root of the R-Square value, and the *p* value is the same as for the regression between these variables (Output 2.2).

The upper portion of the output, as in Output 2.3, contains the partitioning of the sums of squares and the corresponding mean squares. Other items in the output associated with the circled numbers are explained as follows:

① The F value of 10.56 is used to test the null hypothesis

$$H_0: \beta_1 = \beta_2 = \beta_3 = \beta_4 = 0 \quad .$$

The associated p value of < 0.001 leads to the rejection of this hypothesis and indicates that at least one of the coefficients is not zero.[7]

② R-Square = 0.6013 tells you that a major portion (>60%) of the variation of CPM is explained by variation in the independent variables in the model. Note that this R-Square value is much larger than that for the regression using ALF alone (see Output 2.3). The statistic Adj R-Sq = 0.5443 is an alternative to R-Square that is adjusted for the number of parameters in the model according to the formula

Adj R-Sq$=1 - \{(1 - $ R-Square$)* [(n-1)/(n-m-1)]\}$

where n is the number of observations in the data set, and m is the number of regression parameters in the model excluding the intercept. This adjustment provides for measuring the reduction in the mean square (rather than sum of squares) due to the regression. It is used to overcome an objection to R-Square as a measure of goodness of fit of the model, namely, that R-Square can be driven to one (suggesting a perfect fit) simply by adding superfluous variables to the model with no real improvement in the fit. This is not the case with Adj R-Sq, which indicates the point of diminishing returns as more variables are added to the model (Rawlings, Pantula, and Dickey 1998, Section 7.5.3). When m/n is small, say less than 0.05, the adjustment almost vanishes. Adj R-Sq can have values less than zero.

③ Rounding the values of the parameter estimates provides the equation for the fitted model

$$\widehat{CPM} = 8.596 - (7.211)(ALF) - (0.213)(UTL) + (0.333)(ASL) - (4.950)(SPA).$$

Thus, for example, a one-unit (percent) increase in the average load factor, ALF, is associated with a decreased cost of 7.211 cents per passenger mile, holding constant all other factors.[8]

④ The estimated standard errors of the parameter estimates are useful for constructing confidence intervals as illustrated in Section 2.2, "A Model with One Independent Variable." Remember, however, to use the degrees of freedom for MSE.

⑤ The t statistics are used for testing hypotheses about the individual parameters. It is important that you clearly understand the interpretation of these tests. This can be explained in terms of comparing the fit of unrestricted and restricted models (see Section 1.1.2). The unrestricted model for all of these tests contains all the variables on the right-hand side of the MODEL statement. The restricted model contains all these variables except the one being tested. Thus, the t statistic of -5.461 for testing the null hypothesis that $\beta_1 = 0$ (there is no effect due to ALF) is actually testing whether the unrestricted four-variable model fits the data better than the restricted three-variable model, which omits ALF. In other words, the test indicates whether there is variation in CPM due to ALF over and above that due to UTL, ASL, and SPA.

---

[7] All tests in PROC REG are constructed with the principle of holding constant all parameters not mentioned in the hypothesis. Thus, this hypothesis holds constant the intercept, a fact that is usually assumed. See the discussion of Type I and Type II sums of squares in Section 2.4.2, "SS1 and SS2: Two Types of Sums of Squares."

[8] This coefficient is quite different from that obtained in the one-variable regression (see Output 2.2). This is why simple two-variable plots or correlations are often not useful in trying to determine the effects of individual variables in a multiple regression model.

The *p* value for this test is less than 0.0001, indicating quite clearly that there is an effect due to ALF. By contrast, the corresponding *p* value for ALF in the regression where it is the only variable is only 0.0321. Note that for UTL and SPA the *p* value for the parameter in the multiple-variable model is smaller than that for the corresponding correlation (see Output 2.4), while for ASL the reverse is true. This type of phenomenon is the result of correlations among the independent variables. This is more fully discussed in the presentation of the Type I and Type II sums of squares (see Section 2.4.2), and in greater detail in Chapter 4, "Multicollinearity: Detection and Remedial Measures."

2.4 Various MODEL Statement Options

This section presents a number of MODEL statement options that produce additional results and modify the regression model. Although these options provide useful results, not all are useful for all analyses. They should be requested only if needed.

2.4.1 The P, CLM, and CLI Options

One of the most common objectives of regression analysis is to compute predicted values

$$\hat{y} = \hat{\beta}_0 + \hat{\beta}_1 x_1 + \ldots + \hat{\beta}_m x_m$$

as well as their standard errors for some selected values of $x_1, \ldots, x_m$. This can be done in several ways with PROC REG. The most direct way to obtain these values is to use the P and CLM options (which may be used singly) in the MODEL statement as follows:

```
proc reg data = air;
    model cpm = alf utl asl spa / p clm ;
    id alf;
run;
```

The results of these computations are printed below the basic PROC REG output (see Output 2.5), and are shown in Output 2.6. Circled numbers identify the specific items in Output 2.6.

Output 2.6
Computing
Predicted
Values

| | | | Output Statistics | | | | |
| | | | ④ Std Error | ⑤ | | ⑥ | |
Obs ①	ALF	② Dep Var CPM	③ Predicted Value	Mean Predict	95%	CL Mean	Residual
1	0.591	2.2580	2.5737	0.1804	2.2042	2.9432	-0.3157
2	0.488	2.2750	2.1362	0.1959	1.7350	2.5373	0.1388
3	0.412	2.3410	3.4398	0.0933	3.2488	3.6309	-1.0988
4	0.397	2.3570	2.4243	0.2604	1.8908	2.9578	-0.0673
5	0.582	2.3630	2.5634	0.1737	2.2077	2.9191	-0.2004
6	0.466	2.4040	2.8794	0.1129	2.6481	3.1107	-0.4754
7	0.535	2.4250	2.2897	0.1859	1.9089	2.6706	0.1353
8	0.434	2.7110	2.7645	0.1402	2.4774	3.0517	-0.0535
9	0.439	2.7430	3.3673	0.0916	3.1796	3.5550	-0.6243
10	0.417	2.7800	2.8730	0.1520	2.5618	3.1843	-0.0930
11	0.4	2.8330	2.6359	0.1926	2.2413	3.0305	0.1971
12	0.41	2.8460	3.1830	0.1483	2.8792	3.4868	-0.3370
13	0.478	2.9060	3.1898	0.0922	3.0009	3.3787	-0.2838
14	0.495	2.9540	2.9325	0.1164	2.6941	3.1708	0.0215
15	0.476	2.9620	2.9747	0.1117	2.7459	3.2034	-0.0127
16	0.539	2.9710	3.0191	0.1397	2.7329	3.3053	-0.0481

Continued

Output 2.6 (Continued) Computing Predicted Values

17	0.409	3.0440	3.3238	0.1237	3.0704	3.5773	-0.2798
18	0.381	3.0960	2.7518	0.1461	2.4525	3.0510	0.3442
19	0.486	3.1400	3.0937	0.1056	2.8774	3.3099	0.0463
20	0.287	3.3060	3.5692	0.2057	3.1477	3.9906	-0.2632
21	0.504	3.3060	2.7484	0.1123	2.5182	2.9785	0.5576
22	0.455	3.3110	3.4825	0.1018	3.2739	3.6912	-0.1715
23	0.405	3.3130	3.2374	0.1444	2.9416	3.5331	0.0756
24	0.422	3.3920	3.4427	0.0937	3.2508	3.6346	-0.0507
25	0.476	3.4370	3.5201	0.1349	3.2438	3.7965	-0.0831
26	0.426	3.4620	3.2451	0.1081	3.0237	3.4665	0.2169
27	0.349	3.5270	3.1495	0.1460	2.8504	3.4485	0.3775
28	0.394	3.6890	3.6437	0.1127	3.4129	3.8745	0.0453
29	0.452	3.7600	3.4879	0.1006	3.2819	3.6939	0.2721
30	0.425	3.8560	3.5653	0.2600	3.0328	4.0978	0.2907
31	0.362	3.9590	3.5205	0.1833	3.1450	3.8960	0.4385
32	0.541	4.0240	3.1584	0.1552	2.8405	3.4762	0.8656
33	0.378	4.7370	4.3018	0.2167	3.8580	4.7457	0.4352

```
                        Sum of Residuals                    0
                 Sum of Squared Residuals              4.35752
                 Predicted Residual SS (PRESS)         5.84594
```

① The ID statement provides for the identification of individual observations, using the values of the variable specified in that statement. In this case, it is the independent variable ALF. (Although one of the independent variables in the model is often used as the ID variable, any other variable in the data set may be used.)

② This column, labeled Dep Var, gives the observed values of the dependent variable, which is CPM. (Compare this with Output 2.1.)

③ This column, labeled Predicted Value, gives results from the P (for predicted) option in the MODEL statement that causes the REG procedure to compute the $\hat{y}$ values corresponding to each observation in the data set.

④ and ⑤ are the results of the CLM option. This option computes the upper and lower 95% confidence limits for the expected value (conditional mean) for each observation. Computation of these limits requires the standard errors of the predicted values, which are listed in the output under number ④. These are computed according to the formula

Std Err Mean Predict = $((\mathbf{x}'(\mathbf{X}'\mathbf{X})^{-1}\mathbf{x})*MSE)**0.5$

where $\mathbf{x}$ is a row vector of the design matrix $\mathbf{X}$ corresponding to a single observation and MSE is the error mean square. This is equal to the square root of the variance of $\hat{y}$ given in Section 1.1.2. The columns headed 95% CL Mean are the upper and lower 95% confidence limits for the mean and are calculated by

$$(\hat{y} \pm t * (\text{Std Err Mean Predict}))$$

where t is the 0.05 level tabulated t value with degrees of freedom equal to those of the error mean square. For example, for observation 1, for which ALF=0.591 and the actual CPM is 2.2580, you can see that the predicted value is 2.5737, and the 95% confidence limits on the mean response are (2.2042 to 2.9432). The 95% confidence level is used by default; other levels may be obtained with the ALPHA = option in the PROC REG statement.

⑥ This column, labeled Residual, is also produced by the P option in the MODEL statement and contains the following set of residual values:

Residual = CPM − Predicted Value.

It is sometimes useful to scan this column of values for relatively large values (see Chapter 3).

One of the useful features of PROC REG (as well as other SAS regression procedures) is that the $\hat{y}$ values can be computed for observations not in the data set from which the regression equation was estimated. Uses for this feature are presented in Section 2.10.

CLI, an additional MODEL statement option, computes the prediction intervals for a single observation. The formula for this is similar to that for the confidence interval for the mean, except that the variance of the predicted value is larger than that for the interval of the mean by the value of the residual mean square. The results of the CLI option are not printed here. However, a plot of prediction intervals is presented in Sections 2.6.1, "High-Resolution Plots," and 2.7, "Creating Data I: the OUTPUT Statement."

The interpretations of the CLM and CLI statistics are sometimes confused. Specifically, CLM yields a confidence interval for the subpopulation mean, and CLI yields a prediction interval for a single unit to be drawn at random from that subpopulation. The CLI limits are always wider than the CLM limits because the CLM limits need accommodate only the variability in the predicted value, whereas the CLI limits must accommodate the variability in the predicted value as well as the variability in the future value of y. This is true even though the same predicted value $(\hat{y})$ is used as the point estimate of the subpopulation mean as well as the predictor of the future value.

To draw an example, suppose you walk into a roomful of people and are challenged to (1) guess the average weight of all the people in the room and (2) guess the weight of one particular person to be chosen at random. You make a visual estimate of, say, 150 lbs. You would guess the average weight to be 150 lbs., and you would also guess the weight of the mystery person to be 150 lbs. But you would have more confidence in your guess of the average weight for the entire roomful than in your guess of the weight of the mystery person.

In the airline cost data, the 95% confidence interval for the mean of the first airline is from 2.2042 to 2.9432 cents per mile (see Output 2.6). This means that you can state with 95% confidence that these limits include the true mean cost per passenger mile of the hypothetical population of airlines having the characteristics of that airline. On the other hand, the CLI option for this airline provides the interval from 1.6852 to 3.4623 cents per mile. This means that there is a 95% probability that a single airline chosen from that population of airlines will have a cost that is within those limits.

The following guidelines can assist you in deciding whether to use CLM or CLI:

❑ Use CLM if you want the limits to show the region that should contain the population regression curve.

❑ Use CLI if you want the limits to show the region that should contain (almost all of) the population of all possible observations.

One final note: the confidence coefficients for CLM and CLI are valid on a single-point basis. Bands that are valid simultaneously for all points require computations not directly available in PROC REG, but which can be computed with the IML procedure.

At the bottom of Output 2.6 there are three statistics that are sometimes useful. These (as identified) are the sum of the computed residuals, the sum of squares of the computed residuals, and the PRESS statistic. The sum of residuals should be zero, and the sum of squares should equal the error sum of squares in the analysis of variance at the top of the PROC REG output. If different results are obtained, there is reason to suspect that there has been excessive roundoff error (see Section 2.9, "Creating Data III: ODS Output"). The PRESS statistic is among the several available for detecting the possibility of outlying or influential observations. If the PRESS statistic is considerably larger than the residual sum of squares, outliers may exist. Further details are presented in Section 3.2, "Outlier Detection."

2.4.2 SS1 and SS2: Two Types of Sums of Squares

PROC REG can compute two types of sums of squares associated with the estimated coefficients in the model. These are referred to as Type I (sequential) and Type II (partial) sums of squares. These are computed by specifying SS1 and SS2 as MODEL statement options. The following statements produce Output 2.7:

```
proc reg data=air;
    model cpm = alf utl asl spa / ss1 ss2;
run;
```

(The partitioning of the sums of squares and the parameter estimates, which appear in Output 2.3, are not affected by these options and are not reproduced.) The requested sums of squares are printed as additional columns in the Parameter Estimates section and are labeled Type I SS and Type II SS.

Output 2.7
Types I and
II Sums of
Squares

Variable	DF	Type I SS	Type II SS
Intercept	1	318.29667	14.10804
ALF	1	1.52682	4.64087
UTL	1	2.29757	1.66383
ASL	1	0.17165	0.52410
SPA	1	2.57512	2.57512

The interpretation of these sums of squares is based on the material in Section 1.1.2. In particular, the following concepts are useful in understanding the different types of sums of squares:

❑   partitioning of sums of squares

❑   complete and restricted models

❑   reduction notation.

The Type I SS are commonly called sequential sums of squares and represent a partitioning of the Model SS into component sums of squares due to each variable as it is added sequentially to the model in the order prescribed in the MODEL statement.

The Type I SS for INTERCEPT is simply $(\Sigma y)^2 / n$, which is commonly called the correction for the mean. The Type I SS for ALF (1.5268) is the Model SS for a regression equation containing only ALF and the intercept. This is easily verified as the Model SS in Output 2.3. The Type I SS for UTL (2.2976) is the reduction in the Error SS due to adding UTL to a model that already contains ALF and the intercept. Since Model SS+Error SS = Total SS, this is also the increase in the Model SS due to the addition of UTL to the model containing ALF.

Equivalent interpretations hold for the Type I SS for ASL and SPA. In general terms, the Type I SS for any particular variable is the reduction in Error SS due to adding that variable to a model that already contains all the variables preceding that variable in the MODEL statement.

Note that the sum of the Type I SS is the overall Model SS:

$$6.5711 = 1.5268 + 2.2976 + 0.1716 + 2.5751.$$

This equation shows the sequential partitioning of the Model SS into Type I components corresponding to the variables as they are added to the model in the order given. In other words, the Type I sums of squares are order dependent; if the variables in the MODEL statement are given in different order, the Type I sums of squares will change.

The Type II sums of squares are commonly called partial sums of squares. The Type II SS for a variable is the reduction in Error SS due to adding that variable to the model that already contains all the other variables in the model list. For a given variable, the Type II SS is equivalent to the Type I SS for that variable if it is the last variable in the model list. This is easily verified by the fact that both types of sums of squares for SPA, which is the last variable in the list, are the same. The Type II sums of squares, therefore, do not depend on the order in which the independent variables are listed in the MODEL statement. Furthermore they do not yield a partitioning of the Model SS unless the independent variables are mutually uncorrelated.

Reduction notation (see Section 1.1.2) provides a convenient device to determine the complete and restricted models that are compared if the corresponding one degree of freedom sum of squares is used as the numerator for an *F* test. In reduction notation the two types of sums of squares are as follows:

Parameter	Type I (Sequential)	Type II (Partial)
ALF	R(ALF|INTERCEPT)	R(ALF|INTERCEPT, UTL,ASL,SPA)
UTL	R(UTL|INTERCEPT,ALF)	R(UTL|INTERCEPT, ALF,ASL,SPA)
ALS	R(ASL|INTERCEPT,ALF,UTL)	R(ASL|INTERCEPT, ALF,UTL,SPA)
SPA	R(SPA|INTERCEPT,ALF,UTL,ASL)	R(SPA|INTERCEPT, ALF,UTL,ASL)

F tests derived by dividing the Type II SS by the error mean square are equivalent to the *t* tests for the parameters provided in the computer output. In fact, the Type II *F* statistic is equal to the square of the *t* statistic.

Tests derived from the Type I SS are convenient to use in model building in those cases where there is a predetermined order for selecting variables. Not only does each Type I SS provide a test for the improvement in the fit of the model when the corresponding term is added to the model, but the additivity of the sums of squares can be used to assess the significance of a model containing, say, the first $k<m$ terms.

Associated with the Type I sums of squares is the MODEL statement option SEQB, which prints the coefficients of the successive models fitted by adding terms in the order given in the MODEL statement. Type I sums of squares and the SEQB option are very useful in fitting polynomial models, which are presented in Section 5.2, "Polynomial Models with One Independent Variable."

PROC REG does not provide the *F* tests associated with the Type I and Type II sums of squares. This causes no difficulty for the Type II sums of squares since these are the squares of the corresponding *t* statistics. *F* ratios corresponding to the Type I sums of squares must therefore be calculated manually. [9]

2.4.3 Standardized Coefficients: the STB Option

The MODEL statement option STB produces the set of standardized regression coefficients. Output 2.8 contains the additional portion of the output from the statements

```
proc reg data = air;
    model cpm = alf utl asl spa / stb;
run;
```

These estimates appear in the same format as the Type I and Type II sums of squares in the Parameter Estimates section of Output 2.5.

Output 2.8
Standardized
Coefficients

```
                            Standardized

        Variable    DF        Estimate

        Intercept   1          0
        ALF         1         -0.82592
        UTL         1         -0.52768
        ASL         1          0.37011
        SPA         1         -0.75983
```

These coefficients, labeled Standardized Estimate, are the estimates that would be obtained if all variables in the model were standardized to zero mean and unit variance prior to performing the regression computations. Each coefficient indicates the number of standard deviation changes in the dependent variable associated with a standard deviation change in the independent variable, holding constant all other variables. In other words, the magnitudes of the standardized coefficients are not affected by the scales of measurement of the various model variables and thus may be useful in ascertaining the relative importance of the effects of independent variables not affected by the scales of measurement.

2.4.4 Printing Matrices: the XPX and I Options

The XPX and I options are available for printing the matrices used in the regression computations (see Sections 1.1.1, 1.1.2, and 1.3). The XPX option, standing for X prime X, prints the following matrix:

$$\begin{bmatrix} \mathbf{X'X} & \mathbf{X'Y} \\ \mathbf{Y'X} & \mathbf{Y'Y} \end{bmatrix}$$

In other words, it is the matrix of sums of squares and crossproducts of all the variables in the MODEL statement. The I option, standing for Inverse, prints the following matrix:

$$\begin{bmatrix} (\mathbf{X'X})^{-1} & \hat{\beta} \\ (\hat{\beta})' & \text{ERROR SS} \end{bmatrix}$$

---

[9] The *F* tests for Type I sums of squares are provided by PROC GLM, which is, however, not normally used for ordinary regressions.

The output (not including the usual regression printout) from the following statements appears in Output 2.9:

```
proc reg data=air;
    model cpm = alf utl asl spa / xpx i;
run;
```

Output 2.9
Results of
PROC REG
with XPX
and I
Options

```
                                    The REG Procedure
                                     Model: MODEL1

                            Model Crossproducts X'X X'Y Y'Y

Variable        Intercept          ALF          UTL          ASL          SPA          CPM

Intercept              33       14.711       287.67       48.479       6.0581      102.488
ALF                14.711     6.701339     127.6024    21.489977    2.6055474    45.220067
UTL                287.67     127.6024    2574.8875     441.54459    54.159475    883.33621
ASL                48.479    21.489977    441.54459    84.737677    10.0323764   146.297362
SPA                6.0581     2.6055474    54.159475    10.0323764    1.36961579   18.3154399
CPM               102.488    45.220067    883.33621    146.297362   18.3154399   329.225344

                      X'X Inverse, Parameter Estimates, and SSE

Variable        Intercept          ALF          UTL          ASL          SPA          CPM

Intercept   5.2369479797  -6.606496223  -0.264323466  0.5549364635  -4.208572139  8.5955250498
ALF         -6.606496223  11.205647867  0.1644020761  -0.640165877  6.0925283374  -7.211373248
UTL         -0.264323466  0.1644020761  0.0272207146  -0.047184442  0.1256213977  -0.212815723
ASL         0.5549364635  -0.640165877  -0.047184442  0.2112878183  -0.918587822  0.3327689316
SPA         -4.208572139  6.0925283374  0.1256213977  -0.918587822  9.5162522535  -4.950301373
CPM         8.5955250498  -7.211373248  -0.212815723  0.3327689316  -4.950301373  4.3575188862
```

The rows and columns of the matrices are identified by the names of the variables in the model. The variable INTERCEPT corresponds to the dummy variable whose value is unity for all observations and which is used to estimate the intercept coefficient (β_0). Thus in the **X′X** matrix the first row (or column) consists of the sample size and the sums of the variables. All other elements are (uncorrected) sums of squares and crossproducts of the variables.

The rows and columns of the inverse corresponding to INTERCEPT and the independent variables is the inverse of **X′X**. The row and column corresponding to the dependent variable contain the estimated coefficients (compare with Parameter Estimates in Output 2.5). The last element, which has both row and column identified by the name of the dependent variable, contains the error (residual) sum of squares. If there are several dependent variables, then the appropriate number of rows and columns corresponding to the dependent variables containing the coefficient estimates and error sums of squares will be printed.

Additional options COVB and CORRB (not illustrated here) print the matrix of variances and covariances and the matrix of correlations of the estimated coefficients, respectively. The ALL option prints all of the statistics corresponding to the various model options plus some descriptive statistics of the model variables.

2.4.5 Regression through the Origin: the NOINT Option

The MODEL option NOINT forces the regression response to pass through the origin; that is, the estimated value of the dependent variable is forced to be zero when all independent variables have the value zero. An example of this requirement occurs in some growth models where the response (say, weight) must be zero at the beginning, that is, when time is zero. In many applications this requirement is not reasonable, especially when this condition does not or cannot actually occur. If this option is used in such situations, the results of the regression analysis are often grossly misleading. Moreover, even when the conditions implied by the NOINT option are reasonable, the results of the analysis have some features that may mislead the unwary practitioner.

The NOINT option is illustrated with the airline data. This example demonstrates the uselessness of this option when it is not justified since, in this example, zero values of the variables in the model cannot occur. Use PROC REG with the following statements:

```
proc reg data=air;
    model cpm = alf utl asl spa / noint p;
run;
```

Note that in addition to the NOINT option, the P option is added to obtain the predicted values. The results appear in Output 2.10.

Output 2.10
Results of
PROC REG
with NOINT
and P
Options

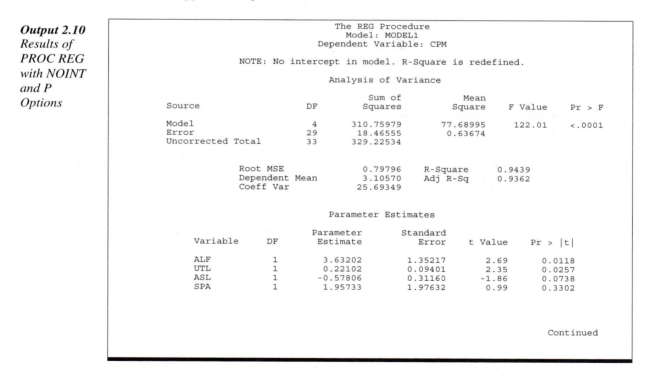

```
                            The REG Procedure
                             Model: MODEL1
                         Dependent Variable: CPM

           NOTE: No intercept in model. R-Square is redefined.

                          Analysis of Variance

                                  Sum of         Mean
     Source             DF        Squares       Square    F Value   Pr > F

     Model               4      310.75979      77.68995    122.01   <.0001
     Error              29       18.46555       0.63674
     Uncorrected Total  33      329.22534


               Root MSE              0.79796   R-Square    0.9439
               Dependent Mean        3.10570   Adj R-Sq    0.9362
               Coeff Var            25.69349


                         Parameter Estimates

                          Parameter     Standard
          Variable   DF    Estimate        Error    t Value   Pr > |t|

          ALF         1     3.63202      1.35217       2.69     0.0118
          UTL         1     0.22102      0.09401       2.35     0.0257
          ASL         1    -0.57806      0.31160      -1.86     0.0738
          SPA         1     1.95733      1.97632       0.99     0.3302




                                                            Continued
```

*Output 2.10
(Continued)
Results of
PROC REG
with NOINT
and P
Options*

```
                                      Output Statistics

                               Dep Var      Predicted
                   Obs             CPM          Value        Residual

                     1          2.2580         3.1204         -0.8624
                     2          2.2750         3.1124         -0.8374
                     3          2.3410         2.8401         -0.4991
                     4          2.3570         2.9600         -0.6030
                     5          2.3630         3.1237         -0.7607
                     6          2.4040         3.4065         -1.0025
                     7          2.4250         3.6856         -1.2606
                     8          2.7110         2.9350         -0.2240
                     9          2.7430         2.8566         -0.1136
                    10          2.7800         2.7355          0.0445
                    11          2.8330         3.1537         -0.3207
                    12          2.8460         3.4019         -0.5559
                    13          2.9060         3.0813         -0.1753
                    14          2.9540         3.3919         -0.4379
                    15          2.9620         3.3846         -0.4226
                    16          2.9710         3.1119         -0.1409
                    17          3.0440         3.2583         -0.2143
                    18          3.0960         3.2487         -0.1527
                    19          3.1400         3.2980         -0.1580
                    20          3.3060         2.6366          0.6694
                    21          3.3060         3.3730         -0.0670
                    22          3.3110         2.8790          0.4320
                    23          3.3130         3.3594         -0.0464
                    24          3.3920         3.0495          0.3425
                    25          3.4370         2.7413          0.6957
                    26          3.4620         3.0421          0.4199
                    27          3.5270         2.7570          0.7700
                    28          3.6890         2.9286          0.7604
                    29          3.7600         2.8861          0.8739
                    30          3.8560         2.5304          1.3256
                    31          3.9590         3.0898          0.8692
                    32          4.0240         3.0684          0.9556
                    33          4.7370         2.3996          2.3374


               Sum of Residuals                            1.64132
               Sum of Squared Residuals                   18.46555
               Predicted Residual SS (PRESS)              25.43858
```

The overall partitioning of the sum of squares and the R-Square statistic certainly suggest a well-fitting model. In fact, both the model F and the R-Square statistics are much larger than those for the model with the intercept (see Output 2.5). However, closer examination of the results shows that the error sum of squares for the no-intercept model is actually larger than the *total* sum of squares for the model with intercept.

This apparent contradiction can be explained by remembering what the F and R-Square statistics represent. Specifically, R-Square is defined as (Model SS)/(Total SS), where the Model SS measures the reduction in the sum of squares due to fitting the regression. When fitting a regression model with an intercept, Total SS is the corrected or centered sum of squares for the response variable, which can be interpreted as the Error SS for the model $y = \mu + \varepsilon$. The Model SS shows the reduction from that sum of squares due to fitting the regression model. In this example,

   Total SS = 10.92867, Model SS = 6.57115; hence R-Square = 0.6013.

However, when you are fitting a regression without an intercept, Total SS is the uncorrected or raw sum of squares, which can be interpreted as the Error SS for fitting the model $y = 0 + \varepsilon$, which is obviously a very poor model! The Model SS, then, is the reduction from that sum of squares due to fitting the regression model. In this example,

   Total SS = 329.22534, Model SS = 310.75979; hence R-Square = 0.9439.

Thus, although the error sum of squares for the no-intercept model is much larger than that for the model with intercept (18.46555 versus 4.35782), the R-Square is larger because the reduction in the sum of squares starts from a much larger value. And because the R-Square and F statistics are closely related, a similar argument holds for the F statistic.

These differences are noted in the output as Uncorrected Total for the total sums of squares of the no-intercept model and as Corrected Total for the total sums of squares of the intercept model. Furthermore, both total and error degrees of freedom are larger by one in the no-intercept model since one fewer parameter (the mean) has been estimated. Finally, the statement at the top of the output:

```
NOTE: No intercept term in model. R-Square is redefined.
```

is intended to alert users to the special interpretation of this statistic and the fact that this value may bear no relationship to the value obtained if the intercept is included. A comprehensive discussion of the appropriate uses of R-Square is found in Kwålseth (1985).

Looking further you can also see that the coefficient estimates for the no-intercept model bear no resemblance to those of the intercept model because you have really fitted two entirely different models. Finally, the statistics that follow the listing of predicted and residual values show that the sum of residuals is not zero. This is in contrast to models with intercepts for which, by definition, the sum of residuals is equal to zero. It is characteristic of no-intercept models that the sum of residuals is not zero, although if the true intercept is near zero, the sum of residuals may be quite close to zero.

A simple example is used to illustrate the fact that the use of the NOINT option can result in misleading results even in cases where the true intercept is near zero. Eight data points are generated using the model *y=x*, with a normally distributed error having zero mean and unit variance. Output 2.11 shows the data.

Output 2.11
Data for
NOINT
Option

Obs	X	Y
1	1	-0.35
2	2	2.79
3	3	1.81
4	4	2.00
5	5	3.88
6	6	6.79
7	7	7.67
8	8	6.79

PROC REG is used for models with and without intercept as follows:

```
proc reg data = noint;
   model y = x / P;
   model y = x / noint p;
run;
```

The results of these analyses appear in Outputs 2.12 and 2.13.

Output 2.12
Results of
PROC REG
without
NOINT
Option

```
                               The REG Procedure
                                Model: MODEL1
                             Dependent Variable: Y

                              Analysis of Variance

                                    Sum of        Mean
    Source              DF         Squares       Square     F Value    Pr > F

    Model                1        49.50857     49.50857       34.23    0.0011
    Error                6         8.67758      1.44626
    Corrected Total      7        58.18615


              Root MSE               1.20261     R-Square     0.8509
              Dependent Mean         3.92250     Adj R-Sq     0.8260
              Coeff Var             30.65919

                             Parameter Estimates

                          Parameter     Standard
    Variable     DF        Estimate        Error    t Value    Pr > |t|

    Intercept     1        -0.96321      0.93706      -1.03      0.3436
    X             1         1.08571      0.18557       5.85      0.0011
                               The REG Procedure
                                Model: MODEL1
                             Dependent Variable: Y

                              Output Statistics

                          Dep Var     Predicted
                 Obs            Y         Value     Residual

                   1      -0.3500        0.1225      -0.4725
                   2       2.7900        1.2082       1.5818
                   3       1.8100        2.2939      -0.4839
                   4       2.0000        3.3796      -1.3796
                   5       3.8800        4.4654      -0.5854
                   6       6.7900        5.5511       1.2389
                   7       7.6700        6.6368       1.0332
                   8       6.7900        7.7225      -0.9325


            Sum of Residuals                           0
            Sum of Squared Residuals             8.67758
            Predicted Residual SS (PRESS)       15.57627
```

Output 2.13
Results of
PROC REG
with NOINT
Option

```
                            The REG Procedure
                             Model: MODEL2
                          Dependent Variable: Y

           NOTE: No intercept in model. R-Square is redefined.

                           Analysis of Variance

                                   Sum of          Mean
    Source                  DF     Squares        Square    F Value   Pr > F

    Model                    1   171.06851     171.06851     117.33   <.0001
    Error                    7    10.20569       1.45796
    Uncorrected Total        8   181.27420


              Root MSE               1.20746   R-Square     0.9437
              Dependent Mean         3.92250   Adj R-Sq     0.9357
              Coeff Var             30.78288

                          Parameter Estimates

                         Parameter     Standard
      Variable    DF      Estimate        Error    t Value   Pr > |t|

      X            1       0.91574      0.08454      10.83     <.0001

                          Output Statistics

                        Dep Var    Predicted
                 Obs          Y        Value     Residual

                   1    -0.3500       0.9157      -1.2657
                   2     2.7900       1.8315       0.9585
                   3     1.8100       2.7472      -0.9372
                   4     2.0000       3.6629      -1.6629
                   5     3.8800       4.5787      -0.6987
                   6     6.7900       5.4944       1.2956
                   7     7.6700       6.4101       1.2599
                   8     6.7900       7.3259      -0.5359


         Sum of Residuals                        -1.58647
         Sum of Squared Residuals                10.20569
         Predicted Residual SS (PRESS)           13.25937
```

You can see immediately that both the model F and the R-Square values are much larger for the no-intercept model whereas the residual mean squares are almost identical. Actually both models have estimated very similar regression lines since the estimated intercept (-0.963) is sufficiently close to zero that a hypothesis test of a zero intercept cannot be rejected ($p=0.3436$). However, the residual mean square of the no-intercept model is still somewhat larger and the sum of residuals for the no-intercept model is -1.586 even though the intercept is quite close to zero.

The difference between the test statistics for the models is based on which null hypothesis is tested in each case. Remember that the test for the model is based on the difference between the error and total sum of squares. The error sum of squares measures the variability from the estimated regression while the total sum of squares measures the variability from the model specified by the null hypothesis.

In the model with an intercept, the null hypothesis $\beta_1 = 0$ specifies the model

$$y = \beta_0 + \varepsilon = \mu + \varepsilon \quad .$$

Now μ is estimated by the sample mean. Therefore, the test for the model compares the variation from the regression to the variation from the sample mean.

In the no-intercept model, the null hypothesis $\beta_1 = 0$ specifies the model $y = \varepsilon$, that is, a *regression* where the mean response is zero for all observations. The test for this model then compares the variation from the regression to the variation from $y = 0$. Obviously, unless the mean of the response variable is close to zero, the variation from the mean is smaller than the variation from zero, hence the apparent contradiction.

You can readily make this comparison from the above example. Since the true intercept is indeed zero, the error sums of squares are not very different (8.68 with intercept, 10.21 without intercept), but the total sums of squares are much larger for the no-intercept model (181.27 for no intercept, and 58.19 with intercept). Finally, since R-Square is based on the difference between the error and the total sum of squares as defined for the particular model, this statistic is also much larger for the no-intercept model.

2.5 Further Examination of Model Parameters

In addition to testing the statistical significance of the model and the individual parameters, you can investigate other properties of the model. Specifically, you can

❏   test hypotheses on any linear function(s) of the parameters. This is performed by the TEST statement described in Section 2.5.1, "Tests for Subsets and Linear Functions of Parameters."

❏   change the model by adding or deleting variables. This can be done by specifying a new model or, interactively, by using ADD and DELETE statements as described in Section 2.5.2. Additional model selection methods are presented in Chapter 4.

❏   examine the effect on the model parameters when some observations are deleted. This is performed by using the REWEIGHT statement described in Section 2.5.2.

❏   apply restrictions to linear functions of parameter estimates. This is done by using the RESTRICT statement described in Section 2.5.3, "Restricted Least Squares."

Before continuing it is important to note that most of the analyses presented in the next several chapters are of an *exploratory* nature. That is, having proposed a model and performed the regression analysis you may also want to consider some special hypotheses or determine the effects of various modifications of either model or data. Often such investigations will involve tests of hypothesis tests and result in *p*-values. It is important to remember that any particular *p*-value refers only to the probability of erroneously rejecting that particular null hypothesis. Obviously, as you continue to make additional hypothesis tests, the overall *experimentwise* probability of making a Type I error increases. Therefore, the *p*-values resulting from such analyses cannot be taken too literally; instead they should be regarded as relative indicators of statistical significance.

2.5.1 Tests for Subsets and Linear Functions of Parameters

The *t* tests on the parameters in the basic PROC REG output (see Output 2.5) provide the tests of hypotheses that individual regression coefficients are zero. Section 2.4.2 indicates that the Type I sums of squares may be used to test whether certain subsets of coefficients are zero. This section shows how the TEST statement can be used to test the hypothesis that one or more linear functions of parameters are equal to specified constants.

One or more TEST statements may accompany a MODEL statement in a PROC REG step. These statements may be entered interactively. The general form of a TEST statement is

   label : **TEST** *equation* <, . . ., *equation>*;

The label is any valid SAS name and serves to identify different tests in the output. The equations specify the linear functions to be tested.

The tests can be interpreted in terms of comparing full (or unrestricted) and restricted models in the manner described in previous sections. The unrestricted model for all tests specified by a TEST statement is the model containing all variables on the right-hand side of the MODEL statement. The restricted model is derived by imposing the restrictions implied by the equations specified in the TEST statement. For this example the following statements are added to the MODEL statement in Section 2.3:

```
test1 : test spa = 0, alf = 0;
test2 : test utl = 0, alf = 0;
test3 : test utl - alf = 0;
```

The TEST1 statement tests the hypothesis that the coefficients for both SPA and ALF are zero. This test compares the unrestricted model with one containing only UTL and ASL. The TEST2 statement tests the hypothesis that the coefficients for both UTL and ALF are zero. This test compares the unrestricted model with the one containing only ASL and SPA. The TEST3 statement tests the hypothesis that the difference between the coefficients for UTL and ALF is zero. This is equivalent to the test that the coefficient for UTL is equal to the coefficient for ALF. (This particular test has no practical interpretation in this example and is used for illustration only.)

A useful feature of these statements is that if the value of the function to be tested is zero, the equality need not be specified. In other words, the TEST1 statement can also be written

```
test1 : spa, alf;
```

The results of these TEST statements appear in Output 2.14.

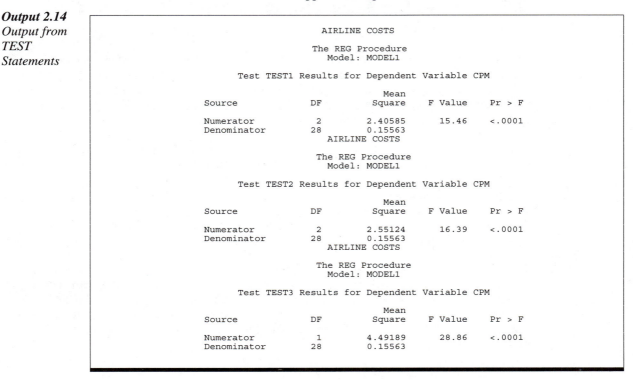

```
                              AIRLINE COSTS

                            The REG Procedure
                             Model: MODEL1

          Test TEST1 Results for Dependent Variable CPM

                                   Mean
     Source              DF       Square     F Value    Pr > F

     Numerator            2      2.40585      15.46     <.0001
     Denominator         28      0.15563
                              AIRLINE COSTS

                            The REG Procedure
                             Model: MODEL1

          Test TEST2 Results for Dependent Variable CPM

                                   Mean
     Source              DF       Square     F Value    Pr > F

     Numerator            2      2.55124      16.39     <.0001
     Denominator         28      0.15563
                              AIRLINE COSTS

                            The REG Procedure
                             Model: MODEL1

          Test TEST3 Results for Dependent Variable CPM

                                   Mean
     Source              DF       Square     F Value    Pr > F

     Numerator            1      4.49189      28.86     <.0001
     Denominator         28      0.15563
```

For each indicated TEST statement, a partial (Type II) sum of squares is computed with degrees of freedom equal to the number of equations in the TEST statement. A mean square is then computed that forms the numerator of an F statistic. The denominator of this statistic is the error mean square for the unrestricted model. The values of the two mean squares, the degrees of freedom, the F ratio, and its p value are printed in the output.

Each of these tests is statistically significant (all p-values <0.0001), which means that deleting SPA and ALF, or deleting UTL and ALF, creates a significant loss of fit. Finally, you cannot conclude that the coefficients for UTL and ALF are equal.

Note: If there are linear dependencies or inconsistencies among the equations of a TEST statement, PROC REG prints a message that the test failed and no F ratio is computed.

It is again important to note that each test is performed independently rather than simultaneously with the other tests. Therefore, the results of the various tests are usually not independent of each other. Also, in making multiple hypothesis tests, you run an increased risk of making an experimentwise Type I error.

2.5.2 Modifying the Model or the Data

Modifying the Model. Although the test for the null hypothesis that ASL and UTL are zero was rejected (Output 2.14), you may be still interested in examining a model that omits these two variables.

You can, of course, simply submit a new PROC REG with the desired model. However, if you are in an interactive mode and have not entered a QUIT command, you have two choices:

❑ Enter a new MODEL statement with any options that may be useful.

❑ Use the interactive DELETE and ADD statements.

Before continuing with examples, you need to understand how PROC REG implements multiple models. PROC REG examines all MODEL statements preceding a RUN statement and computes a matrix of sums of squares and crossproducts of all variables involved in all of those MODEL statements. For all subsequent models, PROC REG uses elements of this matrix to obtain estimates. Therefore, <u>any subsequent model may contain any subset of these but may not contain variables not included in that matrix</u>.

Following the PROC REG statement with a VAR (or VARIABLES) statement listing all variables that you may want to consider for a PROC REG session modifies this procedure by causing PROC REG to compute the sums of squares and crossproducts of all variables listed in the MODEL and VAR statements. Subsequent models can now be analyzed with any subset of these variables without regard to the specification of the initial models.

One by-product of this process is that any observation with a missing value in any one of the variables in the VAR statement or initial set of MODEL statements is not used in the computation of the matrix of sums of squares and crossproducts. Therefore, it is possible that some observations will not be used for some subset models for which the variables are available. Therefore, if missing values are a problem, you should use separate PROC REG steps for each model.

The interactive ADD and DELETE statements are of the form

ADD variables;

and

DELETE variables;

Assuming that you have used the original four-variable model for the airline cost data, you can delete variables ASL and UTL by entering the following statement:

```
delete asl utl;
```

The SAS log will show the following notations:

```
ASL has been deleted.
UTL has been deleted.
```

No output is produced since you may want to perform additional interactive modifications. The output is obtained by issuing the following statement:

```
print;
```

Because PROC REG is still working on the original model, the output will include all MODEL options originally specified for that model. Since some options (especially those presented in Chapter 3) produce rather voluminous output, you may want to use the statement:

```
print anova;
```

which will cause the printing of only the basic PROC REG output, which is shown in Output 2.15.

Output 2.15
Output from
DELETE
Statement

```
                        The REG Procedure
                         Model: MODEL1.1
                     Dependent Variable: CPM

                      Analysis of Variance

                              Sum of         Mean
    Source            DF     Squares        Square    F Value    Pr > F

    Model              2     4.89725       2.44862      12.18    0.0001
    Error             30     6.03142       0.20105
    Corrected Total   32    10.92867


              Root MSE              0.44838    R-Square    0.4481
              Dependent Mean        3.10570    Adj R-Sq    0.4113
              Coeff Var            14.43744


                      Parameter Estimates

                     Parameter     Standard
    Variable     DF   Estimate       Error    t Value    Pr > |t|

    Intercept     1    6.55600      0.72271      9.07      <.0001
    ALF           1   -6.02514      1.36286     -4.42      0.0001
    SPA           1   -4.16369      1.01692     -4.09      0.0003
```

Additional ADD and DELETE statements may follow any ADD or DELETE statement. All such statements are cumulative in that they start from the previous statements. Also remember that additions are restricted to those variables for which the sums of squares and crossproducts have been computed.

Modifying the Data. The REWEIGHT statement interactively changes the weights of observations that are used in computing the sums of squares and crossproducts (see Section 3.4, "Heterogeneous Variances"). This statement is of the form

    REWEIGHT   *<condition>* / *weight = < >;*

Condition identifies the observations to be reweighted, remembering that the default weight of each observation is initially unity. If the weight specification is omitted, the weight is zero, which means the specified observations are deleted. For example, you can estimate the regression model without using the data from airlines with CPM>4.0 as follows:

```
reweight  cpm>4;
print anova;
run;
```

The SAS log will identify the deleted observations, numbers 32 and 33 (Output 2.1) in this example. The regression with the reweighted observations is refitted by the PRINT command. Output 2.16 shows the results.

Output 2.16
Deleting
Observations
with the
REWEIGHT
Statement

```
                              The REG Procedure
                               Model: MODEL1.1
                            Dependent Variable: CPM

                              Weight: REWEIGHT

                            Analysis of Variance

                                  Sum of          Mean
  Source                DF        Squares        Square     F Value    Pr > F

  Model                  4        4.14318       1.03579        8.77    0.0001
  Error                 26        3.07137       0.11813
  Corrected Total       30        7.21455


                  Root MSE              0.34370    R-Square     0.5743
                  Dependent Mean        3.02345    Adj R-Sq     0.5088
                  Coeff Var            11.36780


                            Parameter Estimates

                           Parameter      Standard
  Variable      DF          Estimate         Error    t Value   Pr > |t|

  Intercept      1           7.65260       0.92278       8.29    <.0001
  ALF            1          -6.90821       1.25766      -5.49    <.0001
  UTL            1          -0.12506       0.06548      -1.91    0.0672
  ASL            1           0.25766       0.16344       1.58    0.1270
  SPA            1          -4.42400       1.10141      -4.02    0.0004
```

You can see that the regression estimates are not changed much by the deletion of these observations. A residual plot for this regression, using the same specifications as the previous plot, can be produced by the PLOT statement (see Section 2.6, "Plotting Results") but is not reproduced here.

Other weights may be specified by an option, for example:[10]

```
reweight cpm > 4 / weight = xx;
```

REWEIGHT statements are cumulative and may be undone by the following statement:

```
reweight allobs / reset;
```

2.5.3 Restricted Least Squares

Restricted least squares are used to place linear restrictions on the estimated regression parameters. For example, a restriction may specify the value of a coefficient to be zero, which is the same as deleting the coefficient. Another restriction may specify that the sum of coefficients is some constant.

PROC REG allows restricted least squares estimation with the use of a RESTRICT statement. The RESTRICT statement follows a MODEL statement and has the general form

RESTRICT *equation*<, . . ., *equation*>;

---

[10] For most general applications, weighting that involves all observations is more efficiently done by the WEIGHT statement presented in Section 3.4.1, "Weighted Least Squares."

where each equation is a linear combination of model parameters set equal to a constant. As with the TEST statement, no constant is needed if the restriction equals zero.

The RESTRICT statement is interactive, but successively implemented RESTRICT statements are not cumulative in that they restrict from the original unrestricted model. However, if a RESTRICT statement is followed by a DELETE statement, the deletion is from the restricted model.

The no-intercept model is actually a special case of restricted least squares models. That is, you can fit a no-intercept model by adding the following statement instead of using the NOINT option in a MODEL statement:

```
restrict intercept = 0;
```

The output from the RESTRICT statement with the data set in Output 2.11 is given in Output 2.17. It is instructive to compare the results of the restriction to the NOINT model in Output 2.13, since both are fitting the same model.

Output 2.17
PROC REG
with
RESTRICT
Statement

```
                              The REG Procedure
                                Model: MODEL1
                            Dependent Variable: Y

NOTE: Restrictions have been applied to parameter estimates.

NOTE: Restrictions on intercept. R-Square is redefined.


                            Analysis of Variance

                                    Sum of          Mean
      Source              DF       Squares         Square    F Value    Pr > F

      Model                1     171.06851      171.06851     117.33    <.0001
      Error                7      10.20569        1.45796
      Uncorrected Total    8     181.27420


               Root MSE              1.20746     R-Square     0.9437
               Dependent Mean        3.92250     Adj R-Sq     0.9357
               Coeff Var            30.78288


                            Parameter Estimates

                         Parameter      Standard
      Variable     DF     Estimate         Error    t Value    Pr > |t|

      Intercept     1   -4.6404E-16             0     -Infty     <.0001
      X             1      0.91574       0.08454      10.83      <.0001
      RESTRICT     -1     -1.58647       1.54963      -1.02      0.3436*

              * Probability computed using beta distribution.
```

You can see that the two methods do give the same error mean square and estimated coefficient and test for the coefficient. The parameter estimate for the variable denoted RESTRICT pertains to the Lagrangian parameters that are incorporated in the restricted minimization of the error sum of squares. The *p*-value for the restriction is the same as for the intercept in the model with that parameter.

2.6 Plotting Results

An extremely useful part of a regression analysis is the plotting of various results associated with the analysis. Three different approaches for such plotting provided by PROC REG are

❑ high-resolution plots for most of the result of the analysis.[11]

❑ line-printer plots. These are, of course, not as attractive or precise as the high-resolution plots, but they are somewhat easier to use and also provide some special options not available for the high-resolution plots.

❑ the creation of SAS data sets containing results of the analysis. These data sets can then be used for various other SAS procedures, including plots, charts, data summaries, and so on.

2.6.1 High-Resolution Plots

At any stage in the execution of PROC REG you can obtain scatter plots for any pair of variables in the data set involved in the analysis as well as various statistics produced by the regression analysis. PLOT statements similar to the ones used for PROC GPLOT accomplish this. Such plots can involve any of the variables in the data set as well as various statistics produced by the regression analysis. The data variables are identified by their names, and statistics produced by PROC REG are specified by a keyword followed by a period. For example,

P. (or PREDICTED.)   specifies the predicted values.

R. (or RESIDUAL.)   specifies the residual values.

The most frequently used plots are those involving the residuals. The statements

```
plot r.*p./ vref=0 ; run;
```

following the MODEL statement for the four-variable regression (Output 2.5) produce a plot of the residuals against the predicted values. The VREF = 0 option specifies that a horizontal reference line be drawn on the zero value of the vertical axis. The resulting plot is shown in Output 2.18.

---

[11] High-resolution plots require that SAS/GRAPH software be installed on your computer.

Output 2.18
Residual
Plot

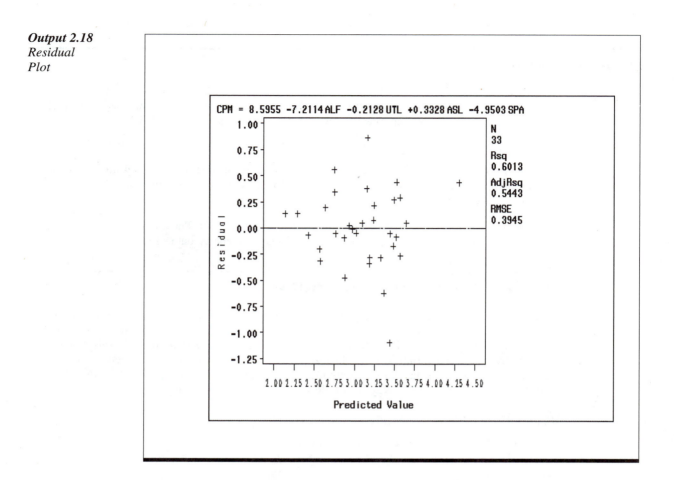

By default, the estimated equation and some summary statistics are shown in the margins. Various options provide for suppressing and/or showing several of the statistics for the regression model. Other options are available for colors, fonts, and symbols, and the plot can be saved to a catalog for PROC GREPLAY.

Plotting options CONF or PRED provide for the plotting of the predicted line and the 0.95 confidence or prediction intervals, respectively, for a regression with one independent variable. Adding the statements

```
plot cpm*alf / pred ; run;
```

after the MODEL statement for the regression of ALF on CPM shown in Output 2.3 produces the plot shown in Output 2.19. On a color printer the lines will be blue, red, and green starting at the top.

Output 2.19
*Prediction
Plot*

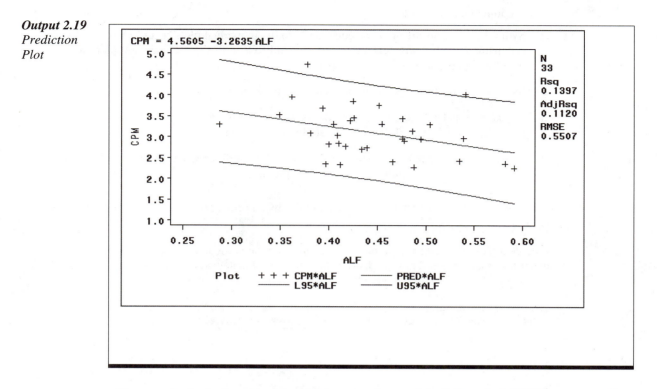

You can obtain similar plots for multiple regression models. Follow the MODEL statement that produces Output 2.5 with the following statements:

```
symbol1 v=plus c=black;
symbol2 v=u c=black;
symbol3 c=l c=black;
plot p.*p. ucl*p. lcl*p. / overlay;
```

The resulting plot is shown in Output 2.20.

Output 2.20
*Prediction
Interval Plot*

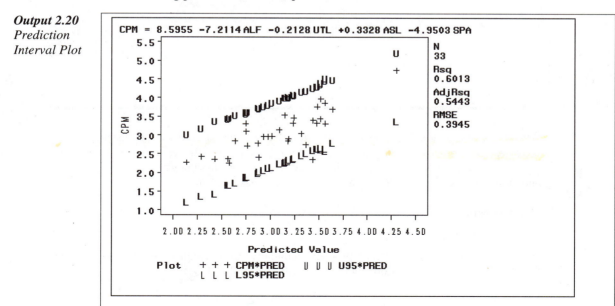

A few comments are in order:

❑ Symbol statements are applied sequentially to the individual plots specified in the PLOT statement. (Symbol numbers are not used as in PROC GPLOT.) The default plot symbol is the plus sign (+) with different colors used to distinguish individual plots.

❑ Symbol options for lines are not available for these plots.

❑ The keywords UCL. and LCL. provide for the prediction interval plots; keywords UCLM. and LCLM. provide the confidence interval plots.

❑ The 95% confidence level is used by default. Other confidence levels may be obtained by specifying ALPHA=*number* as a MODEL option.

You can check on the distribution of the residuals by a normal probability plot. The statements

```
plot r.*nqq.; run;
```

provide a plot of the ordered cumulative residuals on the vertical axis and the corresponding quantiles of the standard normal distribution on the horizontal axis. This plot is also called a Q-Q plot and is shown for the four-variable model in Output 2.21.

Output 2.21
Normal
Probability
Plot

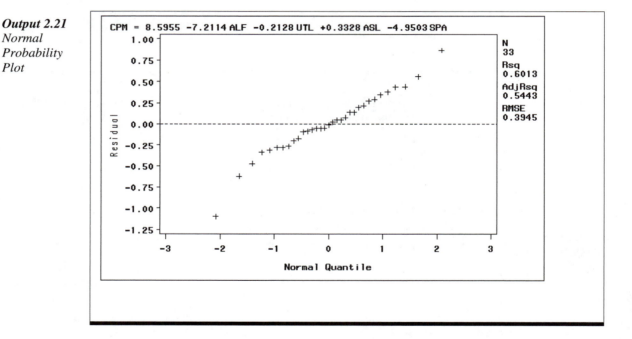

If the points are reasonably close to a straight line it may be inferred that the distribution of residuals is close to normal. If the extreme points deviate vertically, as they are slightly tending to do here, a heavy-tailed distribution is indicated. In this case these deviations are of insufficient magnitude to infer a major violation of the normality assumption.

The statements

```
plot r.*npp; run;
```

provide a so-called p-p plot, which plots the cumulative ranks against the cumulative normal distribution. The function of this plot is similar to that of the Q-Q plot.

A number of other high-resolution plots are available with PROC REG. These are presented at appropriate sections in this book.

2.6.2 Line Printer Plots

Line printer plots are the type of plots created by PROC PLOT. These plots normally use a fixed-pitch font such as Courier or SAS Monospace to create the elements of the plot. They require fewer computer resources than do the high-resolution plots and hence are faster to produce. However, because of the limitations of fixed-pitch fonts, they are not as precise and not as attractive. Line printer plots are available with PROC REG by using the PROC (not MODEL statement) option LINEPRINTER, which specifies that *all* plots for *all* models in that execution of PROC REG will be line printer plots.

Line printer plots with PROC REG are implemented in the same manner as the high-resolution plots but have some options that are not available with high-resolution plots. These include

❏ multiple plots per page. For example, the statement

```
plot a*b c*d e*f g*h / hplots=2 vplots=2;
```

produces four plots on a page: two horizontally (HPLOTS=2) across and two vertically down (VPLOTS=2). If the plot statement specifies only one or two plots, you will get a half page with one or two quarter-page plots.

❏ specifying plotting symbols. By default, the plotting symbol gives the number of replicates at each point; the asterisk symbol (*) is used if there are more than nine replicates. Other symbols are implemented in a manner similar to that of PROC PLOT and are illustrated below.

❏ the PAINT statement, which allows you to highlight or "paint" specified observations in a plot.

❏ the COLLECT option, which allows you to overlay plots from different models.

❏ the VREF and HREF options, which are not implemented.

❏ the OVERLAY option.

Assuming you are using PROC REG with the LINEPRINTER option specified, the following statements after the MODEL statement for the four-variable airline cost data produce two plots:

```
plot r.*p. r.*spa = type / hplots=2 vplots=2;
run;
```

The resulting plots are shown in Output 2.22.

Output 2.22
Line Printer
Residual
Plots

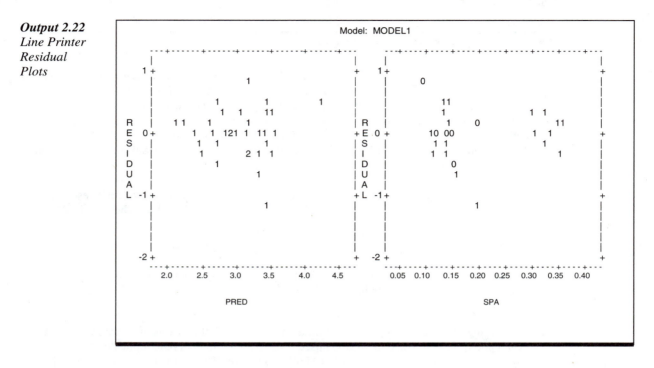

The first plot is for the residual values (R.) on the vertical axis and predicted values (P.) on the horizontal axis. The default plotting symbol indicates the number of observations for each point. The second plot is for the residual values on the vertical axis and the variable SPA on the horizontal axis, using the value of TYPE as the plotting symbol.[12] Remember, TYPE is a binary variable where the value zero indicates short-haul lines with average stage length less than 1200 miles, and unity indicates long-haul lines. The options HPLOTS=2 and VPLOTS=2 provide for two plots both vertically and horizontally, but because there are only two plots there are two half-page plots.

The residuals appear adequately random, but using TYPE as the plotting symbol clearly shows two groups of airlines with respect to average plane size (SPA). As expected, the long-haul airlines tend to have larger planes.

The listing of the data in Output 2.1 shows two airlines with CPM values exceeding 4.0. You can redo the above plots with a highlighting or painting of the residuals for these two airlines with the following statements:

```
paint cpm > 4;
plot;
run;
```

The PAINT statement specifies that all observations for which CPM is larger than 4.0 will be highlighted by the default symbol @. The PLOT statement, without further specification, requests the same plot specifications. The results appear in Output 2.23.

---

[12]   The PLOT statement actually causes PROC REG to refit the latest model, including the effects of any interactive statements preceding the PLOT statement.

Output 2.23
Painting a
Residual
Plot

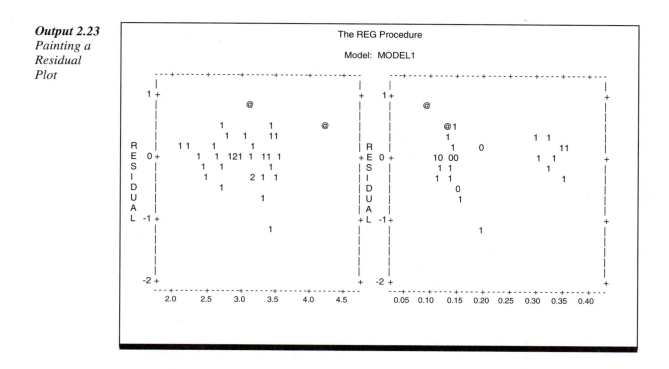

A different symbol for painted observations is specified with the SYMBOL option:

```
paint cpm < 2.5 / symbol = 'S';
plot;
run;
```

This statement paints observations with CPM<2.5 using the symbol S. As PAINT statements are cumulative, the above statement would paint the high-cost airlines with @ and the low-cost airlines with S. A PAINT statement is canceled by the following statement:

```
paint undo;
```

You can overlay plots from different regressions with the COLLECT option. For example, you can examine the effect of a regression using only ALF on the residual plots given in Output 2.22. First, you must return to the original unrestricted model. This can be done by simply repeating the original MODEL statement or by various UNDO and RESET options. Then specify the plots:

```
plot r.*p.= '4' r.*spa= '4' / collect hplots=2 vplots=2;
run;
```

The plotting symbol '4' is specified to indicate that these residuals are from the four-variable model (output not reproduced here). The COLLECT option causes the plot to be saved so that any plots to be specified later can overlay it. Next, delete the variables UTL, ASL, and SPA:

```
delete utl asl spa;
```

At this point you may specify the following statements to examine the results of fitting the resulting model, which is not reproduced here.

```
print anova; run;
```

Now specify the plots:

```
plot r.*p.='1' r.*spa='1'/ nocollect;
run;
```

Note that the plotting symbol '1' is used to show that these residuals are from the one-variable model. The results are shown in Output 2.24.

Output 2.24
Using the
Collect
Option

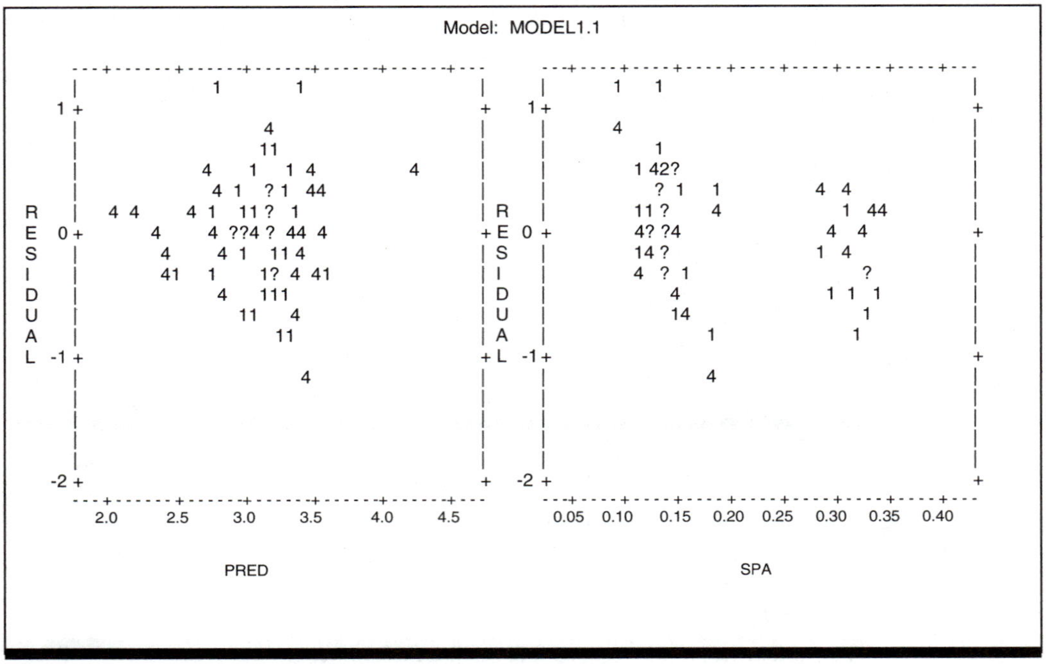

The question mark (?) symbols are data points with multiple symbol specifications. By and large, the residuals from the one-variable model are somewhat larger in magnitude, but details are difficult to ascertain. This shows that overlaid plots are not always useful, because of the lack of definition in a printer plot.

You can use the COLLECT option to produce side-by-side plots with no overlays if you indicate a blank plot for the same position of subsequent collected plots. You must specify a blank (' ') for the plotting symbol. This is illustrated with the plot of residual against predicted values for the same two models: the unrestricted model and the model using only ALF. Use the following statements:

```
model cpm = alf utl asl spa;
plot r.*p.= '4' / hplots=2 vplots=2 collect;
delete utl asl spa;
plot r.*p.=' ' r.*p.='1' / nocollect;
run;
```

The first PLOT statement produces the residual plot for the unrestricted model (not reproduced here). The DELETE statement refits the one-variable model with ALF. The second PLOT statement refits the model and requests two residual plots. However, since the first has blanks as the plotting symbol, nothing is laid over the first plots. The resulting plots appear in Output 2.25.

Output 2.25
Side-by-Side
Residual
Plots

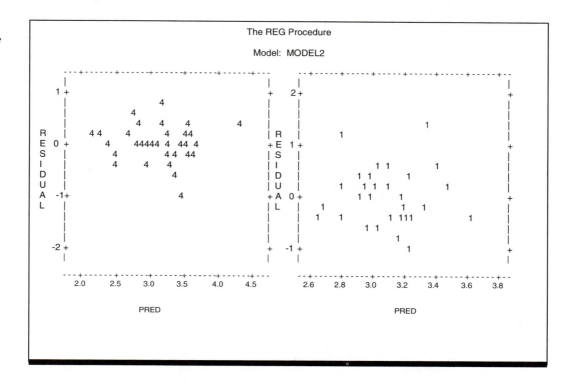

You can now see that the residuals have changed (note the different scales), which was expected with the deletion of three variables.

Here are some additional notes on the COLLECT option:

❑ If the second PLOT statement specifies only one plot, the first plot is overlaid and the second remains as is.

❑ There is no rescaling of axes for the additional collected plots. This may cause loss of plotted points. For this reason you may want to start the collection process with plots showing the largest variation of plotted variables.

❑ Overlaid plots may be collected.

❑ Collection of plots is turned off by using a NOCOLLECT option in the last plot to be collected.

❑ Collection of plots may also be done in order to compare results from REWEIGHT statements.

You can continue with any number of additional interactive statements. However, you must remember that they are cumulative, and after too many such statements you may have forgotten where you are.

2.7 Creating Data I: the OUTPUT Statement

An OUTPUT statement following any MODEL statement causes a new SAS data set to be constructed that contains all of the variables in the data set to which PROC REG was applied, plus other variables for that model as specified in the OUTPUT statement. The basic form of the OUTPUT statement to create a data set containing, for example, predicted and residual values is

> **OUTPUT** OUT=*dsname*   P=*pvarnames* R=*rvarnames*;

where *dsname* is the name chosen for the new data set, *pvarnames* is a list of names chosen for the variables whose values will be the predicted values, and *rvarnames* is the list of names chosen for the variables whose values are the residuals. The names in the *pvarnames* and *rvarnames* lists must correspond to the list of dependent variables in the relevant MODEL statement. A large variety of other output statistics can be included in the OUTPUT data set. Some are used in the example of this section and others are presented in Chapter 3. The complete set of statistics that can be included is given in the PROC REG chapter in the *SAS/STAT User's Guide*.

The OUTPUT statement produces the statistics only for the estimated model and cannot be altered by any interactive statements. However, the resulting data set can be subject to any modifications allowed by the DATA step and used for any SAS procedures, thus allowing considerable flexibility for summarizing all variables from that data set. Furthermore, the data set may be modified by concatenating or merging with other data sets.

The OUTPUT statement is illustrated using the one-variable regression discussed in Section 2.2 to produce a data set that includes predicted and residual values and the 95% prediction intervals. The SAS statements are

```
proc reg data = air;
    model cpm = alf;
    output out=d p=pcpm r=rcpm ucl=up lcl=low;
run;
```

The OUTPUT statement includes the following options:

OUT=D   specifies that the name of the new data set is D.

P=PCPM   specifies that the predicted values have the variable name PCPM.

R=RCPM   specifies that the residuals have the variable name RCPM.

UCL=UP   specifies that the upper $(1 - \alpha)$ prediction limit has the variable name UP. The default for α is 0.05 and may be altered by the ALPHA= option.

LCL=DOWN   specifies that the lower $(1 - \alpha)$ prediction limit has the variable name DOWN.

Additionally, the options UCLM and LCLM may be employed to provide the upper and lower 95% confidence intervals on the conditional mean (see Section 2.4.1, "The P, CLM, and CLI Options"). Additional available output statistics are presented in Chapter 3.

As a result of the OUTPUT statement, the SAS log contains the following information:

NOTE: Data set WORK.D has 33 observations and 10 variables.

This shows that the procedure has created the new data set: the number of variables comprises the original set of six plus the four additional variables specified in the OUTPUT statement.

The output from PROC REG is the same as in Output 2.3 and is not reproduced here. The data set can now be processed by any SAS procedure. For example, the plot of the actual and predicted values as well as the 95% prediction limits can be provided by SAS/GRAPH software, where the more flexible spacing and interpolation options provide a more useful graph than was available with the PLOT statement.

The required statements are

```
symbol1 v=star c=black;
symbol2 v=P i=join c=black;
symbol3 v=U i=spline c=black;
symbol4 v=L i=spline c=black;
proc gplot data=d;
plot
    cpm*alf=1
    pcpm*alf=2
    up*alf=3
    down*alf=4 / overlay;
run;
```

The resulting graph appears in Output 2.26.

Output 2.26
Prediction
Plot Using
SAS/GRAPH
Software

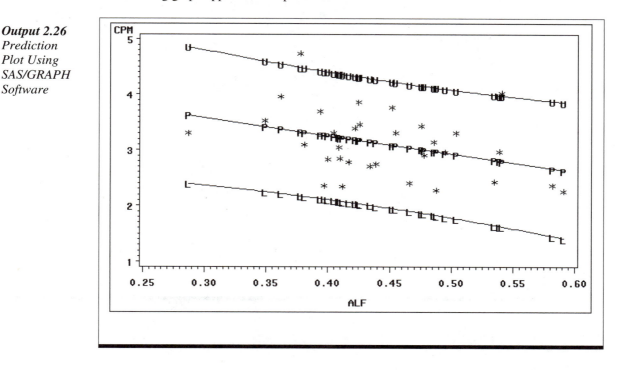

The Normal Q-Q plot shown in Output 2.21 checked the normality of the distribution of the residuals. PROC UNIVARIATE with the PLOT option provides a more complete picture of this distribution:

```
proc univariate plot data=d;

var rcpm; run;
```

The result is shown in Output 2.27.

Output 2.27
PROC
UNIVARIATE
Output for
Residuals

```
                          The UNIVARIATE Procedure
                       Variable:  rcpm  (Residual)
                                  Moments

N                             33      Sum Weights               33
Mean                           0      Sum Observations           0
Std Deviation         0.54204044      Variance          0.29380784
Skewness              0.67082901      Kurtosis          0.51247244
Uncorrected SS        9.40185087      Corrected SS      9.40185087
Coeff Variation                .      Std Error Mean    0.09435713

                      Basic Statistical Measures
            Location                        Variability

     Mean        0.00000       Std Deviation        0.54204
     Median     -0.04510       Variance             0.29381
     Mode              .       Range                2.31799
                               Interquartile Range  0.67657

                    Tests for Location: Mu0=0
         Test              -Statistic-      -----p Value------

         Student's t    t         0      Pr > |t|    1.0000
         Sign           M      -0.5      Pr >= |M|   1.0000
         Signed Rank    S     -19.5      Pr >= |S|   0.7333

                      Quantiles (Definition 5)
                      Quantile        Estimate

                      100% Max        1.4100740
                      99%             1.4100740
                      95%             1.2290324
                      90%             0.6745766
                      75% Q3          0.2917243
                      50% Median     -0.0450983
                      25% Q1         -0.3848495
                      10%            -0.6357337
                      5%             -0.8749653
                      1%             -0.9079186
                      0% Min         -0.9079186

                       The UNIVARIATE Procedure
                     Variable:  rcpm  (Residual)
                          Extreme Observations
         ------Lowest------          -----Highest-----
             Value     Obs              Value     Obs

         -0.907919       7           0.579857       3
         -0.874965      12           0.674577      19
         -0.692936      26           0.682461      15
         -0.635734      21           1.229032      31
         -0.433167      17           1.410074       4

                                                   Continued
```

***Output 2.27
(Continued)
PROC
UNIVARIATE
Output for
Residuals***

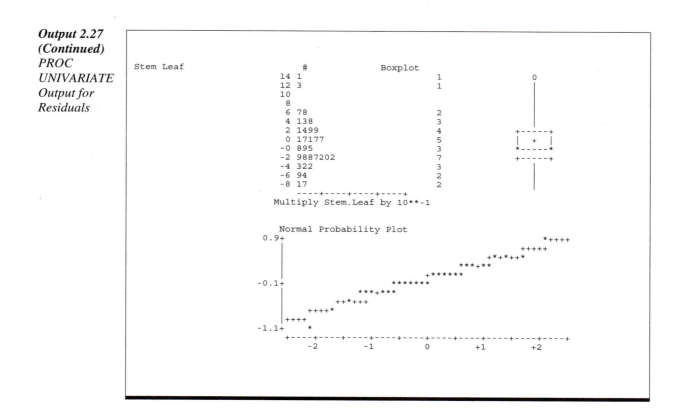

PROC UNIVARIATE provides a rather complete description of the distribution of the variable. For example the means is zero and the sum of squares is indeed Error SS from the regression analysis. Note that the variance is different because PROC UNIVARIATE uses $(n-1)$ for degrees of freedom. The stem and leaf, box, and normal probability plots at the bottom of the output provide information on possible anomalies. There appear to be no problems with these residuals.

2.8 Creating Data II: Other Data Sets

The PROC statement option OUTEST= *dsname* produces a special SAS data set containing estimates and various statistics from the regression model. For each BY group on each dependent variable occurring in each MODEL statement, this option outputs an observation to the OUTEST= data set. The variables output to the data set are as follows:

❑ the BY variables, if any.

❑ _MODEL_, a character variable containing the label of the corresponding MODEL statement, or MODEL*n* if no label is specified, where *n* is 1 for the first model, 2 for the second model, and so on.

❑ _TYPE_, a character variable with the value 'PARMS' for every observation.

❑ _DEPVAR_, the name of the dependent variable.

❑ _RMSE_, the root mean squared error or the estimate of the standard deviation of the error term.

❑ INTERCEPT, the estimated intercept, unless the NOINT option is specified.

❑   all the variables listed in any MODEL or VAR statement. Values of these variables are the estimated regression coefficients for the model. A variable that does not appear in the model corresponding to a given observation has a missing value in that observation. The dependent variable in each model is given a value of -1.

Specifying the PROC statement option TABLEOUT generates additional observations containing the following statistics, identified by the variable _TYPE_ , as follows:

❑   STDERR, the standard error of the estimate

❑   T, the *t* statistic for testing whether the estimate is zero

❑   PVALUE, the associated *p*-value

❑   L*n*B and U*n*B, the lower and upper (1 - α) confidence limits for the estimate, where *n* is the nearest integer to 100*(1 - α) and defaults to 0.05 or is set using the ALPHA= option in the PROC REG statement.

For example, the following statements:

```
proc reg data = air noprint outest=est tableout alpha=.10;
   model cpm = alf utl asl spa;
   proc print data=est;
run;
```

produce the results shown in Output 2.28.

Output 2.28
OUTEST
Data Set

Obs	_MODEL_	_TYPE_	_DEPVAR_	_RMSE_	Intercept	ALF	UTL	ASL	SPA	CPM
1	MODEL1	PARMS	CPM	0.39449	8.5955	-7.21137	-0.21282	0.33277	-4.95030	-1
2	MODEL1	STDERR	CPM	0.39449	0.9028	1.32056	0.06509	0.18133	1.21695	.
3	MODEL1	T	CPM	0.39449	9.5212	-5.46083	-3.26974	1.83512	-4.06779	.
4	MODEL1	PVALUE	CPM	0.39449	0.0000	0.00001	0.00285	0.07713	0.00035	.
5	MODEL1	L90B	CPM	0.39449	7.0598	-9.45782	-0.32354	0.02430	-7.02050	.
6	MODEL1	U90B	CPM	0.39449	10.1313	-4.96492	-0.10210	0.64124	-2.88011	.

Also, if other MODEL statement options, such as variable selection (see Chapter 4) have been specified, the various results of these procedures are also included in the OUTEST data set. The OUTEST data set can be used to predict the response variable using other data sets as shown in Section 2.10. It may also be useful, for example, in summarizing results of a simulation study involving regression.

Additionally, the COVOUT option produces observations containing the covariance matrix of the parameter estimates. The OUTSSCP= option produces a TYPE=SSCP data set containing sums of squares and crossproducts. The SSCP data set can be used as the input to PROC REG, which may be useful when a large number of observations are explored in many different models, as it can be saved and used for subsequent models. This procedure can save considerable computer resources since PROC REG does not need to read the original data again. For details see the PROC REG chapter in the *SAS/STAT User's Guide.*

2.9 Creating Data III: ODS Output

The traditional SAS outputs, consisting of the printed outputs and output data sets, are somewhat limited in their usefulness. SAS outputs can only be printed in line-printer mode and also do not have provisions for extracting specific items for further display or use in analyses. The output data sets do not provide very readable output, are not available for all procedures, and often do not include all results of the procedure.

The SAS Output Delivery System (ODS), available with Version 7 or later of the SAS System, has been designed to overcome the limitations of traditional SAS output. This system provides for outputs to be delivered as a data set, or formatted for high-resolution printer (Postscript), or formatted in Hypertext Markup Language (HTML). These various ODS destinations can further be customized for special uses.

Because traditional SAS output is adequate for almost all applications covered in this book, there is limited presentation of the SAS Output Delivery System. One example is given in Section 3.2.2, "Influence Statistics," for plotting an output statistic that is not available in an output data set. For additional information, see *The Complete Guide to the SAS Output Delivery System, Version 8.*

2.10 Predicting to a Different Set of Data

A regression equation estimated from one set of data can be used to predict values of the dependent variable using the same model for another set of similar data. This type of prediction has been used, for example, to settle charges of pay discrimination against minorities. A regression model relating pay to factors such as education, performance, and tenure is estimated using data for nonminority employees, usually white males. The resulting equation predicts what the pay rate should be for all employees in the absence of discrimination or other factors. If this equation predicts substantially higher than actual salaries for minority employees, there is cause to suspect pay discrimination.

This method may also be used for cross-validation, where a data set is split randomly into two parts and the effectiveness of a statistical procedure used on one part is determined by finding how well it works on the other.

Using a regression equation from one data set to predict the response for another can be accomplished by using the OUTEST option in PROC REG to produce the parameter estimates for the first set, and by using this data set with PROC SCORE to calculate the estimated responses for the second set.

This type of analysis is illustrated with the airline cost data. The variable TYPE was created to divide the airlines into two groups: the short-haul lines with ASL<1200 miles and the long-haul lines with ASL≥1200 miles. One way to ascertain if there are differences in the cost structures between these two types is to see how well the cost equation for short-haul lines estimates costs for the long-haul lines.[13]

---

[13] Another method for answering this question is given in Section 6.4, "Indicator Variables."

First use the IF-THEN/ELSE statements to create two data sets for the long- and short-haul lines:

```
data short long;  set air;
if type = 0 then output short;
else output long;
```

Perform the regression for the short-haul lines and create the OUTEST data set:

```
proc reg data=short outest = shortest;
   model cpm = alf utl asl spa;
run;
```

The regression results are shown in Output 2.29.

Output 2.29
Regression
for Short-
Haul Line

```
                             The REG Procedure
                              Model: MODEL1
                          Dependent Variable: CPM

                            Analysis of Variance

                                   Sum of         Mean
     Source              DF       Squares       Square     F Value    Pr > F

     Model                4       3.90858      0.97714       17.49    0.0003
     Error                9       0.50292      0.05588
     Corrected Total     13       4.41150


               Root MSE              0.23639     R-Square     0.8860
               Dependent Mean        3.33557     Adj R-Sq     0.8353
               Coeff Var             7.08694


                           Parameter Estimates

                           Parameter      Standard
     Variable     DF        Estimate         Error    t Value   Pr > |t|

     Intercept     1        10.70004       1.11685       9.58    <.0001
     ALF           1        -6.72040       1.65219      -4.07    0.0028
     UTL           1        -0.41389       0.05585      -7.41    <.0001
     ASL           1        -0.27371       0.65235      -0.42    0.6846
     SPA           1        -5.49132       3.45925      -1.59    0.1469
```

The regression is certainly significant, but the coefficients are quite different from those obtained from all airlines. The OUTEST data set is now used by PROC SCORE with the following statements:

```
proc score data=long score=shortest out=predlong type=parms
   predict;
   var cpm alf utl asl spa;
run;
```

The PROC options for PROC SCORE are

DATA=LONG   specifies the data set to be used for prediction.

SCORE=SHORTEST   specifies that the OUTEST data set contains the estimated parameters.

OUT=PREDLONG   specifies the output data set that will contain the predicted values.

TYPE=PARMS   is required when using parameter estimates from PROC REG. Different keywords are used for other procedures such as PROC FACTOR.

PREDICT   specifies that predicted values are to be produced. Use RESIDUAL if residual values are wanted.

The data set PREDLONG contains all variables in the original data set plus a variable denoted by MODEL1, which comprises the predicted values. This data set can now be used to plot these predicted against actual values of the response variable:

```
proc plot data=predlong;
plot model1*cpm; run;
```

The results are shown in Output 2.30.

Output 2.30
Plot of
Predicted
Values

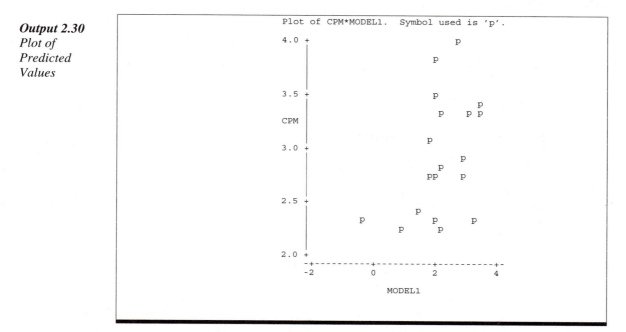

The regression using the short-haul lines obviously does not do a good job of predicting costs for the long-haul lines.

There is another way of predicting responses for observations that are not in the data set being used for the regression. PROC REG will produce predicted values (and prediction or confidence intervals) for observations that have the values of all independent variables but have the dependent variable missing. Thus, if in the DATA step you insert the following:

```
if type = 0 then shortcpm=cpm;
else shortcpm=.;
```

then using PROC REG:

```
proc reg;
model shortcpm = alf utl asl cpm/ p;
```

will produce the same results. Additional steps will be needed to reproduce the plot. This method is primarily useful if predicted values are needed for a few hypothetical observations that can be manually added to the data set.

2.11 Exact Collinearity: Linear Dependency

Linear dependency occurs when exact linear relationships exist among the independent variables. More precisely, a linear dependency exists when one or more columns of the **X** matrix can be expressed as a linear combination of other columns. This means that the **X'X** matrix is singular and cannot be inverted in the usual sense to obtain parameter estimates.[14] PROC REG is programmed to detect the existence of exact collinearity and, if it exists, uses a generalized inverse (see Section 1.1.4, "Using the Generalized Inverse") to compute *parameter estimates*. The words *parameter estimates* are stressed to emphasize that care must be exercised to determine exactly what parameters are being estimated. More technically, the generalized inverse approach yields only one of many possible solutions to the normal equations.

PROC REG computations with an exact linear dependency are illustrated with an alternate model of the airline cost data.

First, create a data set with an additional variable PASS, as follows:

```
data depend; set air;
pass = alf*spa;
```

PASS represents the average number of passengers per flight. Adding this variable does not create a linear dependency since PASS is defined with a multiplicative relationship. Next use a model in which all variables have been converted to logarithms. (The logarithmic model is discussed in some detail in Section 6.2, "Multiplicative Models.") This is done by adding the following statements to the DATA step above:

```
array a alf utl asl spa cpm pass;
array la lalf lutl lasl lspa lcpm lpass;
do over a;
   la = log(a);
end;
```

The ARRAY statements create arrays of the values listed; array A contains the original variables, and array LA is defined to contain the logarithms. The DO OVER statement causes each element of array LA to be the logarithm of the corresponding elements in array A.

Because PASS=ALF*SPA, LPASS=LALF+LSPA, which constitutes an exact collinearity. In the model using the logarithms, then, PASS=ALF+SPA. Now implement PROC REG using all five variables:

```
proc reg data = depend;
   model lcpm = lalf lutl lasl lspa lpass;
run;
```

---

[14] Exact collinearity should not be confused with multicollinearity, which is a term used to describe a high degree of correlation among the independent variables. Multicollinearity can be described as almost exact collinearity. In fact, cases of multicollinearity may be diagnosed as exact collinearity due to computer round off. Thus these are related topics but do, in fact, represent different conditions requiring different methodology. Multicollinearity and associated methodology are discussed in Chapter 4.

The results appear in Output 2.31.

Output 2.31
Exact
Collinearity

```
                               The REG Procedure
                                Model: MODEL1
                           Dependent Variable: LCPM

                             Analysis of Variance

                                   Sum of         Mean
     Source              DF       Squares       Square    F Value    Pr > F

     Model                4       0.70234      0.17558      12.36    <.0001
     Error               28       0.39764      0.01420
     Corrected Total     32       1.09997


               Root MSE              0.11917    R-Square     0.6385
               Dependent Mean        1.11647    Adj R-Sq     0.5869
               Coeff Var            10.67377

 NOTE: Model is not full rank. Least-squares solutions for the parameters are not unique.
 Some statistics will be misleading. A reported DF of 0 or B means that the estimate is
 biased.
 NOTE: The following parameters have been set to 0, since the variables are a linear
 combination of other variables as shown.

                             LPASS =  LALF + LSPA

                             Parameter Estimates

                          Parameter      Standard
     Variable     DF       Estimate         Error    t Value    Pr > |t|

     Intercept    1        0.63879       0.39010       1.64      0.1127
     LALF         B       -1.03377       0.18012      -5.74      <.0001
     LUTL         1       -0.46779       0.15840      -2.95      0.0063
     LASL         1        0.11081       0.08473       1.31      0.2016
     LSPA         B       -0.33717       0.08083      -4.17      0.0003
     LPASS        0              0            .          .         .
```

The existence of the exact collinearity is indicated by the note in the model summary, followed by the equation describing the linear relationship:

LPASS = + 1.0000 * LALF + 1.0000 *L SPA

The parameter estimates that are printed are equivalent to those that would be obtained if LPASS were not included in the MODEL statement. In general, the parameter estimates are those that would be obtained if any variable that is a linear function of variables that precede it in the MODEL statement were deleted from the MODEL statement. These deleted variables are indicated with zeroes under the DF and Parameter Estimate headings. Other variables involved in the linear dependencies are indicated with a B, standing for *bias*, under the DF heading. These estimates are in fact unbiased estimates of the parameters of the model that does not include the deleted variable(s), but are biased estimates for other models.

The bias can readily be seen by using a model with the independent variables listed in a different order. For example, in the model

```
model lcpm = lpass lalf lutl lasl lspa;
```

LSPA will be designated with 0 and LPASS and LALF with B under the DF heading.

Due to round-off error, determining when exact linear dependency occurs is somewhat arbitrary. In PROC REG the matrix inversion procedure computes successive tolerance values. A tolerance of zero signifies an exact collinearity, but since some round-off occurs with virtually all calculations an exact zero tolerance almost never occurs. Therefore a tolerance of 1E-7 is normally used to indicate exact collinearity. This criterion is adequate for virtually all applications, but can be changed by adding the following option to the PROC REG statement:

```
singular = n;
```

where *n* specifies the minimum tolerance for which the matrix is declared nonsingular.

2.12 Summary

The purpose of this chapter has been to provide instruction on how to use PROC REG to perform a regression analysis. This included

❏   estimating the model equation and error variance

❏   providing predicted values and their standard errors

❏   making inferences on the regression parameters

❏   checking the residuals and providing plots to check the fit of the model

❏   making modifications of the model

❏   checking results when some observations have been deleted

❏   finding how well the model may fit an equivalent data set.

Some of the analyses outlined above can be performed in different ways. They also do not exhaust all of the possible analyses that can be performed with PROC REG.

You are now ready to further investigate how well the data fit the model (see Chapter 3) and the suitability of the variables you have chosen for your model (see Chapter 4). Later chapters expand the variety of regression models that you can analyze with the SAS System.

Observations

3.1 Introduction

In the linear model

$$\mathbf{Y} = \mathbf{X}\boldsymbol{\beta} + \boldsymbol{\varepsilon}$$

the elements of the vector ε are the differences between the observed values of the y's and those predicted by the model. These elements comprise the *error term* of the model and are specified to be a set of independently and normally distributed random variables with mean zero and variance σ^2. This error term represents natural variation in the data, but can also be interpreted as the cumulative effect of factors not specified in the model. Often this variation results in errors that, for practical purposes, behave as specified, in which case the use of linear model methodology may be appropriate. However, this is not always the case and, if the assumptions are violated, the resulting analysis may provide results of questionable validity.

Violations of this assumption can occur in many ways. The most frequent occurrences may be categorized as follows:

❑   The data may contain *outliers*, or unusual observations that do not reasonably fit the model.

❑   A *specification error* occurs when the specified model does not contain all of the necessary parameters. This includes the situations where only linear terms have been specified and the true relationships are curvilinear.

❑   The *distribution* of the errors may be distinctly nonnormal; it may be severely skewed or fat tailed.

❑   The errors may exhibit *heteroscedasticity*, that is, the variances are not the same for all observations.

❑   The errors may be *correlated*, which is a phenomenon usually found in time series data, but is not restricted to such situations.

Violations of the assumptions underlying the random errors are often not so severe as to invalidate the analysis, but this is not guaranteed. Therefore, it is useful to examine the data for possible violations and, if violations are found, to employ remedial measures.

Most methods for detecting violations of assumptions are based on the analysis of the estimated errors, which are the *residuals*:

$$\hat{\varepsilon} = y - X\hat{\beta} \quad .$$

However, other statistics may be used. This chapter presents some tools available in PROC REG to detect such violations. Some of these tools involve the estimated residuals, while others are concerned with the behavior of the independent variables. If assumptions fail, alternate methodologies may be used, and some such methods are presented. However, the coverage of detection methodologies and remedial actions presented here is not exhaustive (Belsley, Kuh, and Welsch 1980).

3.2 Outlier Detection

Observations that do not appear to fit the model, often called outliers, can be quite troublesome since they can bias parameter estimates and make the resulting analysis less useful. For this reason it is important to examine the results of a statistical analysis to ascertain if such observations exist. This is especially important in regression analyses where the lack of structure of the independent variables makes detection and identification of outliers more difficult.

It is important to note that observations that may cause misleading results may be unusual with respect to the independent variables, the dependent variable, or both, and that each of these conditions may create different types of results. Furthermore, the identification of unusual observations alone does not provide directions as to what to do with such observations. Any action must be consistent with the purpose of the analysis and conform to good statistical practice.

The following example comes from a study of manpower needs for operating a U.S. Navy Bachelor Officers Quarters (BOQ) (Myers 1990). The observations are records from 25 establishments. The response variable represents the monthly man-hours (MANH) required to operate each establishment, and the independent variables are

OCCUP   average daily occupancy

CHECKIN   monthly average number of check-ins

HOURS   weekly hours of service desk operation

COMMON   square feet of common use area

WINGS   number of building wings

CAP   operational berthing capacity

ROOMS   number of rooms.

The data are shown in Output 3.1.

Output 3.1
BOQ Data

OBS	OCCUP	CHECKIN	HOURS	COMMON	WINGS	CAP	ROOMS	MANH
1	2.00	4.00	4.0	1.26	1	6	6	180.23
2	3.00	1.58	40.0	1.25	1	5	5	182.61
3	5.30	1.67	42.5	7.79	3	25	25	199.92
4	7.00	2.37	168.0	1.00	1	7	8	284.55
5	16.50	8.25	168.0	1.12	2	19	19	267.38
6	16.60	23.78	40.0	1.00	1	13	13	164.38
7	25.89	3.00	40.0	0.00	3	36	36	999.09
8	31.92	40.08	168.0	5.52	6	47	47	931.84
9	39.63	50.86	40.0	27.37	10	77	77	944.21
10	44.42	159.75	168.0	0.60	18	48	48	1103.24
11	54.58	207.08	168.0	7.77	6	66	66	1387.82
12	56.63	373.42	168.0	6.03	4	36	37	1489.50
13	95.00	368.00	168.0	30.26	9	292	196	1845.89
14	96.67	206.67	168.0	17.86	14	120	120	1891.70
15	96.83	677.33	168.0	20.31	10	302	210	1880.84
16	97.33	255.08	168.0	19.00	6	165	130	2268.06
17	102.33	288.83	168.0	21.01	14	131	131	3036.63
18	110.24	410.00	168.0	20.05	12	115	115	2628.32
19	113.88	981.00	168.0	24.48	6	166	179	3559.92
20	134.32	145.82	168.0	25.99	12	192	192	2227.76
21	149.58	233.83	168.0	31.07	14	185	202	3115.29
22	188.74	937.00	168.0	45.44	26	237	237	4804.24
23	274.92	695.25	168.0	46.63	58	363	363	5539.98
24	384.50	1473.66	168.0	7.36	24	540	453	8266.77
25	811.08	714.33	168.0	22.76	17	242	242	3534.49

Perform the regression using the following SAS statements:

```
proc reg data = boq;
    model manh = occup checkin hours common wings cap rooms;
```

The results appear in Output 3.2.

Output 3.2
Regression for
BOQ Data

```
                          The REG Procedure
                           Model: MODEL1
                       Dependent Variable: manh

                         Analysis of Variance

                                  Sum of        Mean
Source                DF         Squares       Square    F Value    Pr > F

Model                  7        87387188     12483884      60.26    <.0001
Error                 17         3522013       207177
Corrected Total       24        90909201

           Root MSE            455.16727    R-Square     0.9613
           Dependent Mean     2109.38640    Adj R-Sq     0.9453
           Coeff Var            21.57818

                         Parameter Estimates

                        Parameter      Standard
Variable      DF         Estimate         Error    t Value    Pr > |t|

Intercept      1        134.96790     237.81430       0.57      0.5778
OCCUP          1         -1.28377       0.80469      -1.60      0.1291
CHECKIN        1          1.80351       0.51624       3.49      0.0028
HOURS          1          0.66915       1.84640       0.36      0.7215
COMMON         1        -21.42263      10.17160      -2.11      0.0504
WINGS          1          5.61923      14.74609       0.38      0.7079
CAP            1        -14.48025       4.22018      -3.43      0.0032
ROOMS          1         29.32475       6.36590       4.61      0.0003
```

The test for the model is certainly significant. However, only three of the coefficients appear to be important, and among these, the effect of CAP has an illogical sign. Such results are typical when multicollinearity exists, which is the topic of Chapter 4, "Multicollinearity: Detection and Remedial Measures." For now, the focus is on finding unusual observations.

3.2.1 Residuals and Studentized Residuals

The traditional tool for detecting outliers (as well as specification errors; see Section 3.3, "Specification Errors") consists of examining the residuals.

A difficulty with residuals is that they are not all estimated with the same precision. However, you can compute standard errors of residuals, and when the residuals are divided by these standard errors, you will obtain standardized or *studentized* residuals. These studentized residuals follow Student's *t* distribution. For error degrees of freedom exceeding ten, values from the *t* distribution greater than 2.5 are relatively rare. Thus, studentized residuals exceeding this value provide a convenient vehicle for identifying unusually large residuals.

As indicated in Chapter 2, "Using the REG Procedure," PROC REG provides these residuals, as well as other statistics used for detecting violations of assumptions, in several ways. These may be

❑ printed as part of the PROC REG output as specified by options in the MODEL statement

❑ plotted with the interactive plotting capabilities of PROC REG

❑ output to a data set for printing, plotting, or other statistical analyses.

For printing these statistics as part of the regression output, you specify the R (for RESIDUAL) option in the MODEL statement:

```
proc reg data = boq;
   model manh = occup checkin hours common wings cap rooms/ r;
```

The additional results produced by this option appear in Output 3.3.

Output 3.3
R Option for
BOQ Data

```
                            The REG Procedure
                             Model: MODEL1
                        Dependent Variable: MANH
                             Output Statistics

        Dep Var  Predicted    Std Error               Std Error  Student                      Cook's
  Obs    MANH      Value    Mean Predict Residual     Residual   Residual   -2-1 0 1 2          D

    1  180.2300  209.9847    230.8767  -29.7547         392.3    -0.0759    |        |        0.000
    2  182.6100  213.7956    182.5672  -31.1856         416.9    -0.0748    |        |        0.000
    3  199.9200  380.7026    174.8009 -180.7826         420.3    -0.430     |        |        0.004
    4  284.5500  360.1059    183.8290  -75.5559         416.4    -0.181     |        |        0.001
    5  267.3800  510.3725    181.4372 -242.9925         417.4    -0.582     |*       |        0.008
    6  164.3800  360.4859    182.8696 -196.1059         416.8    -0.470     |        |        0.005
    7  999.0900  685.1674    194.6507  313.9226         411.4     0.763     |      * |        0.016
    8  931.8400  891.8459    163.8236   39.9941         424.7     0.0942    |        |        0.000
    9  944.2100  815.4661    241.2020  128.7439         386.0     0.334     |        |        0.005
   10     1103      1279      272.7556 -176.0594         364.4    -0.483     |        |        0.016
   11     1388      1398      141.6988   -9.9665         432.5    -0.0230    |        |        0.000
   12     1490      1305      204.7782  184.3228         406.5     0.453     |        |        0.007
   13     1846      1711      336.5583  135.0289         306.4     0.441     |        |        0.029
   14     1892      1973      128.9035  -81.7158         436.5    -0.187     |        |        0.000
   15     1881      2751      275.4770 -870.0699         362.3    -2.401     |****    |        0.417
   16     2268      1632      160.3728  635.9230         426.0     1.493     |      **|        0.039
   17     3037      2210      120.7651  826.4958         438.9     1.883     |     ***|        0.034
   18     2628      2190      134.1111  437.9938         435.0     1.007     |      **|        0.012
   19     3560      4225      339.8854 -665.2110         302.7    -2.197     |****    |        0.761
   20     2228      2699      276.2080 -470.9780         361.8    -1.302     |**      |        0.123
   21     3115      3135      288.7190  -19.6054         351.9    -0.0557    |        |        0.000
   22     4804      4386      304.1426  418.4622         338.6     1.236     |      **|        0.154
   23     5540      5864      403.3733 -323.8936         210.9    -1.536     |***     |        1.079
   24     8267      7854      426.0571  413.2654         160.2     2.580     |     *****        5.889
   25     3534      3695      452.5333 -160.2755         48.896   -3.278     |******  |      115.041

                        Sum of Residuals                    0
                        Sum of Squared Residuals      3522013
                        Predicted Residual SS (PRESS) 213738243
```

The various statistics are identified by column headings:

Dep Var MANH	is the dependent variable, MANH.
Predicted Value	lists the predicted values of MANH.
Std Error Mean Predict	lists the standard errors of the estimated conditional means, which is more precisely explained in Section 2.4.1.
Residual	lists the residuals.
Std Error Residual	lists the standard errors of the residuals.
Student Residual	list the studentized residuals. You can verify that the studentized residuals are the residuals divided by their standard errors.
-2-1 0 1 2	contains a schematic plot of the studentized residuals that shows one asterisk for each 0.5 value of the studentized residuals. Thus, five asterisks correspond to studentized residuals exceeding 2.5 in absolute value.
Cook's D	is the Cook's *D* statistic to be discussed later.

Observation 25 is of particular interest. Note that the residual of -160.3 is neither excessively large compared with the other residuals nor the residual standard deviation (ROOT MSE) of 455. But the standard error of the residual, 48.896, is by far the smallest, thus producing the largest studentized residual among all observations. You can also see that observations 15 and 17 have the largest residuals, while observations 24 and 25 have the largest studentized residuals, both of which exceed 2.5.

A better appreciation of these residuals is afforded by examining plots, which may be obtained interactively, or by creating a data set and implementing the PLOT procedure, which is the method used here.

```
proc reg data=boq;
   model manh = occup checkin hours common wings cap rooms;
   output out = resid p = pman r = rman student = student;
proc plot data = resid hpercent=50 vpercent=50;
   plot rman*pman student*pman / vref = 0;
```

The OUTPUT statement produces a data set called RESID, which contains the additional variables PMAN, RMAN, and STUDENT that correspond to the variables specified by the keywords P, R, and STUDENT and are the predicted, residual, and studentized residual values, respectively. The PROC PLOT statement options HPERCENT=50 and VPERCENT=50 provide for the two half-page side-by-side plots for better comparison. The PLOT statement requests the plots of residuals and studentized residuals against the predicted values, and the option VREF=0 provides for a reference line of zero on the vertical axis. The results appear in Output 3.4.

Output 3.4
Residual and
Studentized
Residual
Plots

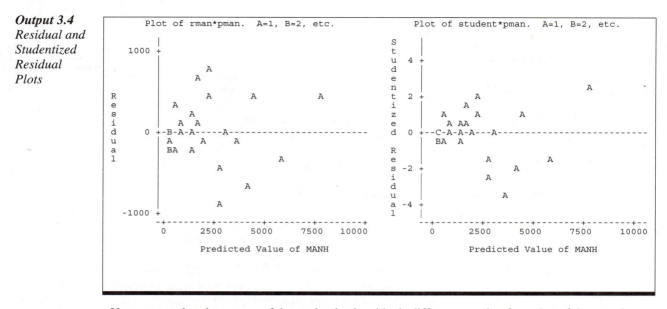

You can see that the pattern of the studentized residuals differs somewhat from that of the actual residuals, and although a number of studentized residuals exceed 2.5 in magnitude, none clearly stand out. The studentized residuals do, however, emphasize that the magnitudes of residuals increase with larger predicted values. This may be an indicator of heterogeneous variances, which are discussed in Section 3.4, "Heterogeneous Variances."

3.2.2 Influence Statistics

Outliers are sometimes not readily detected by an examination of residuals because the least-squares estimation procedure tends to pull the estimated regression response toward observations that have extreme values in either *x* or *y* dimensions. The estimated residuals for such observations may therefore not be especially large, thus hindering the search for outliers. You may be able to overcome this difficulty by asking what would happen to various estimates and statistics if the observation in question were not used in the estimation of the regression equation used to calculate the statistics.

Such statistics are annotated by the subscript $(-i)$ to indicate that they are computed by omitting (subtracting) the *i*th observation. For example, s^2_{-i} is the residual mean square obtained when all except the *i*th observation are used to estimate the regression. Statistics of this type are said to be determining the potential *influence* of a particular observation; hence they are called influence statistics.[1] To print these statistics as part of the PROC REG output, you use the INFLUENCE option as follows:[2]

```
proc reg data = boq;
    model manh = occup checkin hours common wings cap rooms/
    influence;
```

---

[1] The computations of influence statistics do not require the recomputation of regression omitting each observation. All of these statistics are functions of the results of the original model and the hat matrix (see the discussion of the Hat Diag statistic later in this section). Computational and other details can be found in Belsley, Kuh, and Welsch (1980).
[2] You may use the INFLUENCE option in addition to the P and R options (Chapter 2), which is not done here.

The output (not including the default Output 3.2) produced by this option appears in Output 3.5.

Output 3.5
Output Using
the
INFLUENCE
Option

```
                              The REG Procedure
                               Model: MODEL1
                         Dependent Variable: MANH

                             Output Statistics

                                            Hat Diag        Cov
            Obs    Residual     RStudent          H       Ratio      DFFITS

              1    -29.7547      -0.0736     0.2573      2.1809     -0.0433
              2    -31.1856      -0.0726     0.1609      1.9305     -0.0318
              3   -180.7826      -0.4196     0.1475      1.7454     -0.1745
              4    -75.5559      -0.1762     0.1631      1.9109     -0.0778
              5   -242.9925      -0.5704     0.1589      1.6437     -0.2479
              6   -196.1059      -0.4594     0.1614      1.7440     -0.2016
              7    313.9226       0.7532     0.1829      1.5041      0.3563
              8     39.9941       0.0914     0.1295      1.8581      0.0353
              9    128.7439       0.3246     0.2808      2.1428      0.2029
             10   -176.0594      -0.4720     0.3591      2.2688     -0.3533
             11     -9.9665      -0.0224     0.0969      1.7980     -0.0073
             12    184.3228       0.4426     0.2024      1.8475      0.2230
             13    135.0289       0.4299     0.5467      3.2687      0.4722
             14    -81.7158      -0.1818     0.0802      1.7369     -0.0537
             15   -870.0699      -2.8657     0.3663      0.0932     -2.1787
             16    635.9230       1.5537     0.1241      0.6025      0.5849
             17    826.4958       2.0538     0.0704      0.2687      0.5652
             18    437.9938       1.0074     0.0868      1.0874      0.3106
             19   -665.2110      -2.5192     0.5576      0.2536     -2.8282
             20   -470.9780      -1.3310     0.3682      1.1097     -1.0162
             21    -19.6054      -0.0541     0.4024      2.7136     -0.0444
             22    418.4622       1.2566     0.4465      1.3820      1.1286
             23   -323.8936      -1.6057     0.7854      2.2901     -3.0715
             24    413.2654       3.2093     0.8762      0.2461      8.5373
             25   -160.2755      -5.2423     0.9885      0.0473    -48.5179


      --------------------------------DFBETAS------------------------------------
      Obs   Intercept     OCCUP    CHECKIN      HOURS     COMMON      WINGS        CAP      ROOMS

        1    -0.0433   -0.0006    -0.0044     0.0345     0.0036    -0.0020    -0.0007     0.0013
        2    -0.0310   -0.0006    -0.0025     0.0204     0.0046    -0.0020    -0.0008     0.0017
        3    -0.1662    0.0041     0.0037     0.1231    -0.0128     0.0048    -0.0018     0.0020
        4     0.0040    0.0026     0.0103    -0.0504     0.0261    -0.0006     0.0032     0.0003
        5     0.0135    0.0115     0.0517    -0.1597     0.0985    -0.0006     0.0165    -0.0119
        6    -0.1971   -0.0037    -0.0113     0.1318     0.0374    -0.0025     0.0048    -0.0018
        7     0.3234   -0.0274    -0.0700    -0.2195    -0.1375    -0.0160    -0.0576     0.0820
        8    -0.0029   -0.0031    -0.0110     0.0226    -0.0135     0.0020    -0.0033     0.0034
        9     0.1271    0.0070    -0.0108    -0.1376     0.1242    -0.0116     0.0206    -0.0217
       10    -0.0055   -0.0160    -0.0577    -0.1285     0.1666    -0.2715    -0.0783     0.1213
       11     0.0004    0.0010    -0.0002    -0.0045     0.0022     0.0002     0.0013    -0.0007
       12     0.0069    0.0092     0.1441     0.0886    -0.0080     0.0454     0.0129    -0.0671
       13    -0.0413    0.0586    -0.0851     0.0377     0.2267    -0.0205     0.3618    -0.2759
       14     0.0085    0.0068     0.0209    -0.0277     0.0051    -0.0037     0.0106    -0.0113
       15     0.0223   -0.0035    -0.4404    -0.1233    -0.5381    -0.1963    -1.7040     1.4204
       16    -0.0790    0.0289    -0.2033     0.2201     0.1826    -0.2019     0.2104    -0.1149
       17    -0.0929   -0.0984    -0.1058     0.2598     0.0756    -0.0191    -0.1389     0.1178
       18    -0.0243   -0.0054     0.1403     0.1157     0.0965     0.0305    -0.0209    -0.0446
       19    -0.0598    0.9457    -1.7852     0.0088    -0.5855     1.2009     1.3983    -0.9411
       20     0.1531    0.2449     0.7725    -0.1946     0.0217     0.6030     0.4888    -0.6953
       21     0.0065    0.0124     0.0268    -0.0061    -0.0034     0.0275     0.0285    -0.0343
       22    -0.0431   -0.0822     0.7506    -0.1382     0.7150     0.1604    -0.0147    -0.2197
       23     0.2010    0.2396     0.2936     0.3357     0.1652    -2.0577    -0.0658     0.1008
       24     0.7391   -1.6514     0.5958    -1.2140    -5.1382    -0.8793    -0.6539     2.1090
       25    -0.2455  -44.3791     1.1685     0.6569    -4.5712     1.1845    -3.9433     7.2839


                    Sum of Residuals                         0
                    Sum of Squared Residuals           3522013
                    Predicted Residual SS (PRESS)    213738243
```

The various statistics are identified in the column labeled with the keywords used in the OUTPUT and PLOT statements except as noted in the descriptions. The first column in the output repeats the actual residuals. The other columns are as follows:

RStudent

is another version of studentized residuals, where the residuals are divided by a standard error, which uses s^2_{-i} rather than s^2 as the estimate of σ^2. It is thus a more sensitive studentized residual than the studentized residual statistic provided by the R option. If there are no outliers, these statistics should obey a *t* distribution with $(n-m-2)$ degrees of freedom, and therefore the criterion for *large* is the same as for the Student Residual.

Hat Diag H

(keyword H) represents the diagonal elements of the hat matrix $\mathbf{X(X'X)}^{-1}\mathbf{X'}$. The individual values, often denoted by h_i, indicate the *leverage* of each observation, which is a standardized measure of how far an observation is from the center of the space of *x* values. Observations with high leverages, which are indicated by large h_i, have the potential of being influential, especially if they are also outliers in *y*. The sum of h_i is $(m+1)$, where *m* is the number of independent variables; hence the mean value of h_i is approximately $(m+1)/n$, where *n* is the number of observations in the data set. Therefore, observations with $h_i > 2(m+1)/n$ may be considered as having high leverage.

Cov Ratio

(keyword COVRATIO) indicates the change in the precision of the estimates of the set of partial regression coefficients resulting from the deletion of an observation. This precision is measured by the *generalized variance*, which is computed $(s^2)|\mathbf{(X'X)}^{-1}|$. The COVRATIO statistic is the ratio of the generalized variance without the *i*th observation and the generalized variance using all observations, that is:

$$\text{COVRATIO} = \frac{s^2_{-i}\,|\mathbf{(X'X)}^{-1}_{-i}|}{s^2\,|\mathbf{(X'X)}^{-1}|}$$

In other words, COVRATIO values exceeding unity indicate that the inclusion of the observation produces increased precision (smaller variance), while values less than unity indicate decreased precision.[3] Belsley, Kuh, and Welsch (1980) suggest that COVRATIO values outside the bounds defined as $(1 \pm 3(m+1)/n)$ may indicate observations having large influence on the precision.

DFFITS

measures the standardized difference between predicted value for the *i*th observation obtained by the equation estimated by all observations and the equation estimated from all observations except the *i*th, that is, $(\hat{y}_{i,-i} - \hat{y}_i)$. The standardization uses the residual variance estimate from all *other* observations, s^2_{-i}. This statistic is a prime indicator of influence. Belsley, Kuh, and Welsch (1980, p. 28) suggest that DFFITS values exceeding $2[(m+1)/n]^{1/2}$ in absolute value provide a convenient criterion for identifying influential observations.

DFBETAS

is the standardized difference for each individual coefficient estimate resulting from the omission of the *i*th observation. It is identified by column headings with the name of the corresponding independent variable. Since there are many DFBETAS values, it is useful to examine only those for observations with large DFFITS values, where large DFBETAS may indicate which independent variable(s) may be the cause of the influence. DFBETAS values are not available in output data sets or for interactive plotting, but they can be output and plotted by using the ODS statement, as shown later in this section.

---

[3] Relative magnitudes of the determinant of $\mathbf{X'X}$ are indicators of multicollinearity (Chapter 4). Therefore, a COVRATIO statistic also indicates if the deletion of that observation changes the degree of multicollinearity.

A related statistic is Cook's D (keyword COOKD), which is printed by the R option (shown in Output 3.5). Essentially, it is the DFFITS statistic, scaled and squared to make extreme values stand out more clearly. However, the estimated error variance obtained by using all observations is used for standardization, which is the reason it is produced by the R rather than the INFLUENCE option.

Finally, the sum and sum of squares of the actually computed residuals and the Predicted Residual SS (denoted PRESS) are printed at the bottom of the output. The PRESS sum of squares is the sum of squares of residuals using models obtained by estimating the equation with all other observations, that is, $\left(y_i - \hat{y}_{i,-i}\right)$. The sum of residuals should be zero and the sum of squares should be the same as the Error SS in the regression output. Round-off errors may cause minor deviations from these values; larger deviations suggest that severe round-off errors have occurred. The sum of squares of the PRESS statistic should be compared to the Error SS. When it is appreciably larger than the Error SS, as it is here, there is reason to suspect that some influential observations or outliers exist.

The quoted critical values for these statistics will often point to a larger than desired number of suspicious observations, especially with moderate to large data sets. A more satisfactory criterion may be to plot and look for subjectively obvious extreme observations. Creating stem and leaf and box plots with PROC UNIVARIATE or PROC BOXPLOT may also be useful.

The influence statistics now appear to focus more specifically on observations 24 and 25, with 25 having an extremely large DFFITS and COOK'S D (Output 3.3) statistic. The DFBETAS indicate that observation 25 is highly influential with respect to the OCCUP coefficient.

Interactive plots provide a clearer picture. Assuming that PROC REG with the LINEPRINTER option is active (it is not necessary to have specified the INFLUENCE option), enter the following statements:

```
paint obs. = 25;
    plot (student. rstudent. h. dffits.)*p. /
    hplots = 2 vplots = 2;
run;
```

As noted in Section 2.6.2, keywords specifying variables for the PLOT statement are the same as those used in the OUTPUT statement, but each is followed by a period. In this example, all statistics are to be plotted against the predicted values. Because observation 25 appears suspicious, you may highlight it with a PAINT statement, which causes the values for this observation to be plotted with a special symbol. This is implemented by the first statement in the SAS code above, and since no symbol is specified the symbol is the default @. The HPLOTS and VPLOTS options specify that all four plots are on one page.

The results appear in Output 3.6.

Output 3.6
Interactive
Plots for
Influence
Statistics

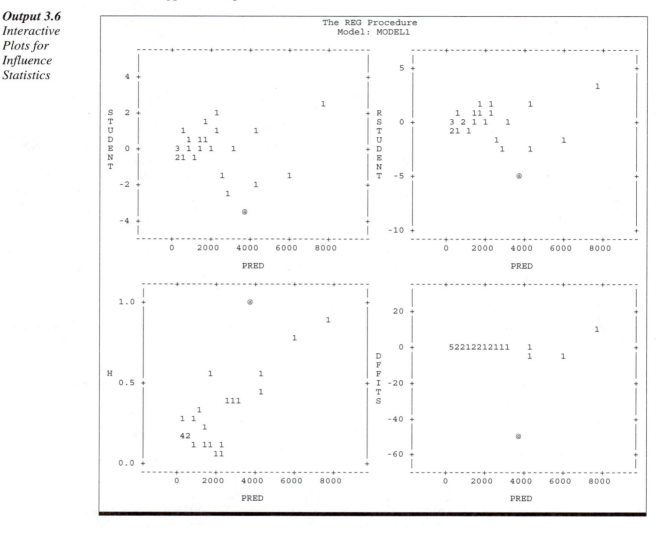

You can see that all of the statistics for observation 25 are large according to the guidelines outlined above, although none except for DFFITS is appreciably larger than the next largest. Further, returning to Output 3.5, the DFBETAS indicate that the OCCUP variable is the culprit. The DFFITS statistic is also quite large for observation 24, and in this case the cause appears to be in the COMMON variable, although the evidence is not very clear.

What to do now is not strictly a statistical problem. *Discarding observations simply because they do not fit is bad statistical practice.* You should first examine the data for obvious discrepancies by various other interactive plots. In addition, you can create an OUTPUT data set and further examine the various statistics with the UNIVARIATE procedure or other descriptive procedures.

In this example, looking at observation 25, and focusing on the OCCUP variable, it becomes obvious that this value is impossible: how can average occupancy be 811 with a capacity of only 242? Observation 24 has a very high CHECKIN rate along with a very small COMMON area and high MANH requirements. One possibility is that this establishment has more transient guests while the others have more longer-term residents.

Normally, an obvious error such as that in observation 25 would be investigated and probably corrected. Observation 24 is, however, probably correct and should be left alone.

Because information on the cause of the obvious error for observation 25 is not available, there is justification for eliminating 25 from the data set and redoing the regression. This can be done in three ways:

❑ by physically eliminating observation 25 from the input file and running PROC REG again

❑ by making the dependent variable (MANH) missing in observation 25, which you can do in the DATA step

```
data miss25; set boq;
   if n = _25_then manh = . ;
run;
```

and then running PROC REG using the DATA= MISS25 statement option

❑ by using the interactive REWEIGHT statement in PROC REG.

As noted in Chapter 2, the REWEIGHT statement assigns an arbitrary weight to any specified observation(s).[4] The interactive statement

```
reweight obs. = 25;
print anova;
```

assigns the (default) zero weight to observation 25. A weight of zero is equivalent to deleting the observation.

The REWEIGHT statement causes a reestimation of the last model, with observation(s) reweighted as instructed, but it creates no output. The PRINT ANOVA statement is used to generate the default output for the reweighted regression because the previous MODEL statement included the INFLUENCE option, which is not needed here.

Output 3.7
Omitting
Observation
25

```
                          The REG Procedure
                           Model: MODEL1.1
                       Dependent Variable: MANH

                           Weight: REWEIGHT

                         Analysis of Variance

                                  Sum of         Mean
     Source            DF         Squares       Square    F Value    Pr > F

     Model              7        87497673     12499668     154.32    <.0001
     Error             16         1295987        80999
     Corrected Total   23        88793659


              Root MSE            284.60353    R-Square     0.9854
              Dependent Mean     2050.00708    Adj R-Sq     0.9790
              Coeff Var            13.88305


                         Parameter Estimates

                         Parameter     Standard
     Variable      DF     Estimate        Error    t Value    Pr > |t|

     Intercept      1     171.47336    148.86168       1.15     0.2663
     OCCUP          1      21.04562      4.28905       4.91     0.0002
     CHECKIN        1       1.42632      0.33071       4.31     0.0005
     HOURS          1      -0.08927      1.16353      -0.08     0.9398
     COMMON         1       7.65033      8.43835       0.91     0.3781
     WINGS          1      -5.30231      9.45276      -0.56     0.5826
     CAP            1      -4.07475      3.30195      -1.23     0.2350
     ROOMS          1       0.33191      6.81399       0.05     0.9618
```

---

[4] Principles of weighted regression are presented in Section 3.4.

The changes in the estimated regression relationship are quite marked. The residual mean square has decreased considerably, the coefficient for OCCUP has become highly significant, while those for CAP and ROOMS are not. Therefore, as indicated by the influence statistics, observation 25 did indeed influence the estimated regression.

Before continuing, it is important to point out that these statistics do not always provide clear evidence of outliers or influential observations. The different statistics are designed to detect different types of data anomalies; hence, they may provide apparently contradictory or confusing results. Furthermore, they may fail entirely, especially if there are several outliers. You can verify this by simply duplicating observation 25 and implementing the INFLUENCE option. In other words, these statistics are only tools to aid in outlier detection; they do not replace careful data monitoring.

Plotting the DFBETAS using ODS Output. As briefly noted in Section 2.9, the SAS Output Delivery System (ODS) provides a means for creating SAS data sets from the output of all SAS procedures. You can use ODS to plot the DFBETAS even though they are not available in a traditional output data set.

The various portions of the output of a SAS procedure are available in different output objects. See the *SAS/STAT User's Guide* for more information on the REG procedure. The following program steps will provide a listing of these objects for the BOQ regression with the INFLUENCE option.

```
ods  trace on;

proc reg data=boq;
   model manh = occup checkin hours common wings cap
   rooms / influence;
run;
ods trace off;
```

Because the ODS output creates objects for all printed output, the PROC REG and MODEL statement must include all options for which you want output. The ODS TRACE ON statement creates a trace record in the SAS log of all available ODS output objects. The ODS TRACE OFF statement, which must follow the RUN statement, cancels this operation. The listing of the files in the log shows the following:

```
Output Added:
-------------
Name:       ANOVA
Label:      Analysis of Variance
Template:   Stat.REG.ANOVA
Path:       Reg.MODEL1.Fit.manh.ANOVA
-------------

Output Added:
-------------
Name:       FitStatistics
Label:      Fit Statistics
Template:   Stat.REG.FitStatistics
Path:       Reg.MODEL1.Fit.manh.FitStatistics
-------------

Output Added:
-------------
Name:       ParameterEstimates
Label:      Parameter Estimates
Template:   Stat.REG.ParameterEstimates
Path:       Reg.MODEL1.Fit.manh.ParameterEstimates
-------------

Output Added:
-------------
Name:       OutputStatistics
Label:      Output Statistics
Template:   Stat.Reg.OutputStatistics
Path:       Reg.MODEL1.ObswiseStats.manh.OutputStatistics
-------------

Output Added:
-------------
Name:       ResidualStatistics
Label:      Residual Statistics
Template:   Stat.Reg.ResidualStatistics
Path:       Reg.MODEL1.ObswiseStats.manh.ResidualStatistics
-------------
```

Note that the table names are also listed in the *SAS/STAT User's Guide.*

The influence statistics are available in the Output Statistics table. You can create the data set A containing these statistics and invoke PROC CONTENTS to examine the contents of the data set as follows:

```
proc reg data=boq;
    model manh = occup checkin hours common wings cap
    rooms / influence;
ods output outputstatistics=a;
proc contents data=a;
run;
```

The relevant portion of the PROC CONTENTS results are shown in Output 3.8.

Output 3.8
Contents of
Output
Statistic
Data Set

```
                              The CONTENTS Procedure

                  -----Alphabetic List of Variables and Attributes-----

       #    Variable          Type     Len    Pos    Format    Label
       -----------------------------------------------------------------------
       7    CovRatio          Num      8      32     8.4       Cov Ratio
       9    DFB_Intercept     Num      8      48     8.4       Intercept DFBETAS
      15    DFB_cap           Num      8      96     8.4       cap DFBETAS
      11    DFB_checkin       Num      8      64     8.4       checkin DFBETAS
      13    DFB_common        Num      8      80     8.4       common DFBETAS
      12    DFB_hours         Num      8      72     8.4       hours DFBETAS
      10    DFB_occup         Num      8      56     8.4       occup DFBETAS
      16    DFB_rooms         Num      8     104     8.4       rooms DFBETAS
      14    DFB_wings         Num      8      88     8.4       wings DFBETAS
       8    DFFITS            Num      8      40     8.4
       2    Dependent         Char    32     144
       6    HatDiagonal       Num      8      24     8.4       Hat Diag H
       1    Model             Char    32     112
       3    Observation       Num      8       0     BEST8.    Obs
       5    RStudent          Num      8      16     D9.3
       4    Residual          Num      8       8     D9.3
```

The output of the PROC CONTENTS shows all of the variables in the data set. All variable names correspond to those in the printed output; the DFBETAS are all named DFB_*name of variable_*. You can merge this data set with the original BOQ data set and then plot or perform other analyses with all these variables. It is useful to plot the DFBETAS against the observation number. Use the following statements:

```
Data b; set a;
   n = _n_;
   symbol1 v='1' c=black;
   symbol2 v='2' c=black;
   symbol3 v='3' c=black;
   symbol4 v='4' c=black;
   symbol5 v='5' c=black;
   symbol6 v='6' c=black;
   symbol7 v='7' c=black;
proc gplot;
   plot dfb_cap*n=1 dfb_checkin*n=2 dfb_common*n=3
   dfb_hours*n=4 dfb_occup*n=5 dfb_rooms*n=6 dfb_wings*n=7
   / overlay;
run;
```

The statement N = _N_ creates the variable *n* as the observation number. The SYMBOL statements assign the symbol 1 to the first listed plot (CHECKIN*N), and so forth. The plot is shown in Output 3.9.

Output 3.9
Plot of
DFBETAS

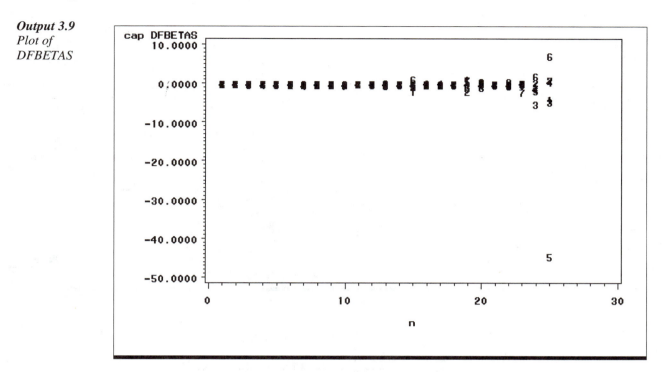

The plot clearly shows that the large DFBETAS values are almost entirely in observation 25, and for that observation the value for OCCUP is by far the largest.

3.3 Specification Errors

Specification error is defined as the result of an incorrectly specified model and often results in biased estimates of parameters. Specification errors may be detected by examining the residuals from a fitted equation. However it is not normally useful to use statistics designed to identify single outliers to detect specification errors, because specification errors are usually evidenced by patterns involving groups of residuals. A common pattern is for the residuals to suggest a curvilinear relationship, although other patterns such as bunching or cycling are possible.

The following example consists of data collected to determine the effect of certain variables on the efficiency of irrigation. The dependent variable is the percent of water percolation (PERC), and the independent variables are as follows:

RATIO   ratio between irrigation time and advance time

INFT   exponent of time in the infiltration equation, a calculated value

LOST   percentage of water lost to deep percolation

ADVT   exponent of time in the water advance equation, a calculated value.

The data appear in Output 3.10.

Output 3.10
Irrigation
Data

Obs	PERC	RATIO	INFT	LOST	ADVT
1	37.75	0.77	0.427	29.10	0.5820
2	34.83	0.97	0.427	29.10	0.6980
3	33.75	1.15	0.309	20.90	0.5190
4	30.26	1.27	0.427	29.10	0.8800
5	30.60	1.27	0.309	20.90	0.6850
6	29.05	1.51	0.309	20.90	0.8000
7	25.76	1.67	0.343	21.27	0.5640
8	26.26	1.87	0.309	20.90	0.8360
9	25.75	2.25	0.309	20.90	0.8410
10	17.16	2.32	0.397	26.53	0.5960
11	14.73	2.34	0.343	21.27	0.7500
12	19.04	2.39	0.427	29.10	0.6000
13	13.16	2.71	0.397	25.45	0.7470
14	18.03	2.82	0.427	29.10	0.7300
15	14.46	3.17	0.397	26.53	0.4360
16	12.96	3.35	0.397	26.53	0.7600
17	16.80	3.64	0.343	21.27	0.6991
18	13.72	3.69	0.387	26.53	0.6000
19	14.56	3.73	0.397	25.45	0.7010
20	12.71	3.78	0.387	25.45	0.6800
21	9.06	4.97	0.397	26.53	0.7720
22	9.52	6.86	0.397	26.53	0.8400

Perform the regression analysis with these SAS statements:

```
proc reg data = irrig;
   model perc = ratio inft lost advt;
   plot r.*p. / vref=0;
run;
```

The regression results appear in Output 3.11 and the residual plot in Output 3.12.

Output 3.11
Regression
for Irrigation
Data

```
                             The REG Procedure
                               Model: MODEL1
                           Dependent Variable: PERC

                            Analysis of Variance

                                   Sum of          Mean
    Source              DF         Squares        Square     F Value    Pr > F

    Model                4      1300.19826     325.04956       16.23    <.0001
    Error               17       340.38669      20.02275
    Corrected Total     21      1640.58495


                  Root MSE              4.47468    R-Square     0.7925
                  Dependent Mean       20.90545    Adj R-Sq     0.7437
                  Coeff Var            21.40436


                           Parameter Estimates

                         Parameter      Standard
    Variable     DF       Estimate         Error    t Value    Pr > |t|

    Intercept     1       40.55825      11.34545       3.57      0.0023
    RATIO         1       -4.82052       0.74296      -6.49      <.0001
    INFT          1     -209.31100     100.93107      -2.07      0.0536
    LOST          1        2.75483       1.35638       2.03      0.0582
    ADVT          1        4.28507       8.82268       0.49      0.6334
```

The regression is certainly statistically significant. The variable RATIO appears to be very important, INFT and LOST contribute marginally, while ADVT appears to have little effect. The plot of residuals is shown in Output 3.12.

Output 3.12
Residual Plot
for Irrigation

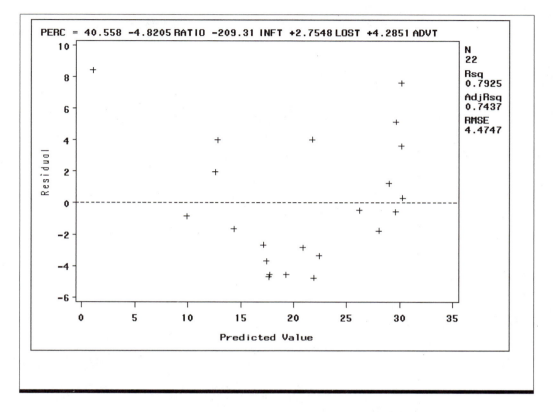

A curved pattern in the residuals is evident. This suggests that a curvilinear component, probably a quadratic term in one or more variables, should be included in the model. Unless there is some prior knowledge to suggest which variable(s) should have the added quadratic, you can use the *partial regression residual plots* for this purpose.

The estimate of the partial regression coefficient β_i can be defined as the coefficient of the simple linear regression of the residuals of y and x_i, respectively, from the regressions involving all other independent variables in the model (Ryan 1997, Section 5.5). A plot of these residuals thus shows the data for estimating that coefficient. If, in this example, such a plot shows a curved pattern, there may be a need for curvilinear terms.[5]

---

[5] These plots may also be useful for outlier detection, which is why they are also called leverage plots. For this purpose, the plots are more useful if an ID statement is used with PROC REG, which causes the first character of the ID variable to be used as the plotting symbol. This feature is primarily useful if the ID variable is a character variable.

Such a plot is called a *partial residual regression* or *partial residual leverage* plot. You can create this plot for RATIO with the following statements:

```
proc reg data = irrig;
   model perc ratio = inft lost advt;
   output out = a r = rperc rratio;
proc gplot data = a;
   plot rperc*rratio / vref=0;
proc reg data = a;
   model rperc = rratio;
run;
```

In this program, the variables RPERC and RRATIO in data set A are the residuals from the regression of PERC and RATIO, respectively, on INFT, LOST, and ADVT. The results of this program provide the following:

❑   the plot of RRATIO against RPERC, which is the partial residual plot and is shown in Output 3.13

❑   the last PROC REG (output not shown), which produces the partial regression coefficient for the RATIO variable in the multiple regression as found in Output 3.11.

Output 3.13
Partial
Residual
Plot for
RATIO

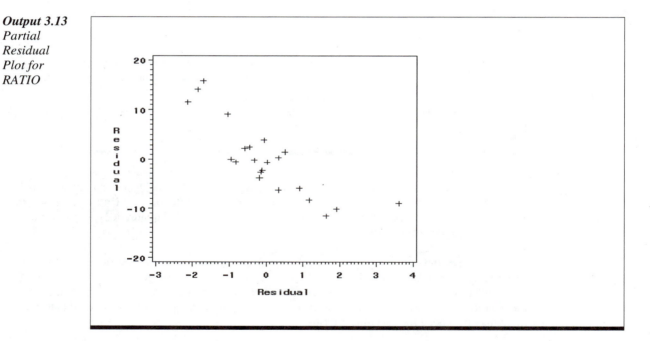

The plot shows a strong negative linear trend with a slight upward curved pattern, suggesting that this variable requires the addition of the quadratic term.

You can avoid all this programming and obtain leverage plots with the MODEL statement option PARTIAL, which produces partial residual plots for all variables. These are always line printer plots and are not reproduced here. The plot involving INFT shows the rather weak negative relationship implied by the p value of 0.536 while none of the other leverage plots reveal anything of interest.

In order to provide for the addition of the quadratic term suggested by the leverage plot, you must create the quadratic value of RATIO in the DATA step:

```
data quad; set irrig;
ratsq = ratio * ratio;
```

Then perform the regression:

```
proc reg data=quad;
   model perc = ratio ratsq inft lost advt;
run;
```

The results appear in Output 3.14.

Output 3.14
Quadratic
Regression
for Irrigation
Data

```
                         The REG Procedure
                          Model: MODEL1
                      Dependent Variable: PERC

                        Analysis of Variance

                              Sum of        Mean
   Source          DF        Squares      Square    F Value   Pr > F

   Model            5     1531.93656   306.38731      45.12   <.0001
   Error           16      108.64838     6.79052
   Corrected Total 21     1640.58495


           Root MSE             2.60586   R-Square     0.9338
           Dependent Mean      20.90545   Adj R-Sq     0.9131
           Coeff Var           12.46499


                        Parameter Estimates

                     Parameter     Standard
   Variable    DF     Estimate        Error   t Value   Pr > |t|

   Intercept    1     53.18731      6.95180      7.65     <.0001
   RATIO        1    -13.05607      1.47466     -8.85     <.0001
   ratsq        1      1.18482      0.20282      5.84     <.0001
   INFT         1   -108.20297     61.27321     -1.77     0.0965
   LOST         1      1.30864      0.82778      1.58     0.1335
   ADVT         1     -0.51707      5.20330     -0.10     0.9221
```

Comparing these results with those of Output 3.11, you can see that the residual mean square has been halved. In other words, a better-fitting equation has been developed. The relationship is one where PERC decreases with RATIO, but the rate of decrease diminishes with increasing values of RATIO. The other variables remain statistically insignificant. Therefore, you may want to consider variable selection (see Chapter 4).

3.4 Heterogeneous Variances

A fundamental assumption underlying linear regression analyses is that all random errors (the ε_i) have the same variance. Outliers may be considered a special case of unequal variances since such observations may be considered to have very large variances.

Residual plots that reveal groupings of observations with large residuals suggesting larger variances usually detect violations of the equal variance assumption. Some of the other outlier detection statistics may also be helpful, especially when the violation occurs in only a small number of observations.

In many applications there is a recognizable pattern for the magnitudes of the variances. The most common of these is an increase in variation for larger values of the response variable. For such cases, the use of a transformation of the dependent variable is in order (Freund and Wilson 1998,

Section 8.2). The most popular of these, especially in regression, is the logarithmic transformation, which is discussed in Chapter 6, "Special Applications of Linear Models."

If the use of transformations is not appropriate, it is sometimes useful to alter the estimation procedure by either modifying the least-squares method or implementing a different estimation principle. A discussion of some alternatives, including additional references, is given in Myers (1990, Section 7.7). An illustration of one of these special methods, using iteratively reweighted least squares, is shown in Freund and Wilson (1998, Section 4.4).

Because the effects of heterogeneous variances are subtle, the effects and remedial methods for this condition are illustrated with a very simple and somewhat pathological example. The data consist of records of sales prices of a set of investment-grade diamonds, and are to be used to estimate the relationship of the sales price (PRICE, in $1000) to weight (CARATS) of the diamonds. The data appear in Output 3.15 and the corresponding SAS/GRAPH plot appears in Output 3.16.[6]

Output 3.15
Diamond
Prices

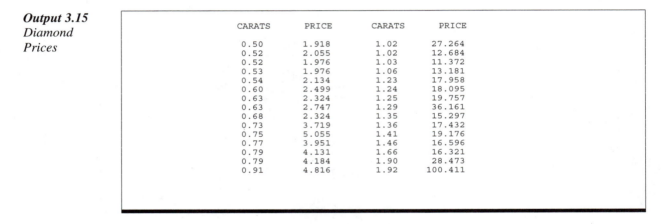

CARATS	PRICE	CARATS	PRICE
0.50	1.918	1.02	27.264
0.52	2.055	1.02	12.684
0.52	1.976	1.03	11.372
0.53	1.976	1.06	13.181
0.54	2.134	1.23	17.958
0.60	2.499	1.24	18.095
0.63	2.324	1.25	19.757
0.63	2.747	1.29	36.161
0.68	2.324	1.35	15.297
0.73	3.719	1.36	17.432
0.75	5.055	1.41	19.176
0.77	3.951	1.46	16.596
0.79	4.131	1.66	16.321
0.79	4.184	1.90	28.473
0.91	4.816	1.92	100.411

Output 3.16
Plot of
Diamond
Prices

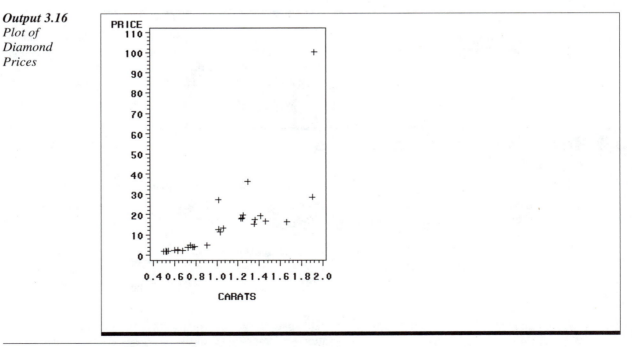

---

[6] To save space the data are presented in two columns using the whole page access of the PUT statement as shown in the *SAS Language: Reference.*

The plot clearly reveals the large variation of prices for larger diamonds. It is difficult to ascertain if the most expensive diamond is an outlier or simply reflects the much higher variability of prices for larger stones. The plot also suggests an upward curving response; hence it is appropriate to specify a quadratic regression. Create the variable CSQ in the DATA step as the square of CARATS, and implement the regression:

```
proc reg data=diamonds;
model price = carats csq / p;
run;
```

The results appear in Output 3.17.

Output 3.17
Diamond
Data,
Ordinary
Least
Squares

```
                              The REG Procedure
                                Model: MODEL1
                          Dependent Variable: PRICE

                             Analysis of Variance

                                    Sum of          Mean
        Source            DF       Squares         Square    F Value    Pr > F

        Model              2    6328.15034     3164.07517      21.59    <.0001
        Error             27    3956.91510      146.55241
        Corrected Total   29        10285


                  Root MSE             12.10588    R-Square     0.6153
                  Dependent Mean       13.86623    Adj R-Sq     0.5868
                  Coeff Var            87.30477


                             Parameter Estimates

                             Parameter      Standard
        Variable      DF      Estimate         Error    t Value    Pr > |t|

        Intercept      1      11.26977      15.30469       0.74      0.4679
        CARATS         1     -30.61470      29.63091      -1.03      0.3107
        CSQ            1      28.43956      12.88434       2.21      0.0360
```

As expected, the regression is certainly significant. The estimated equation is

$$PRICE = 11.27 - 30.614(CARATS) + 28.440(CARATS)^2 \ .$$

The coefficient for $(CARATS)^2$ is statistically significant, confirming the upward curve of the relationship of price to carats. The coefficient for (CARATS) has no practical significance except to indicate that the slope of the curve keeps increasing (see Chapter 5, "Curve Fitting").

The P option to the MODEL statement provides the predicted and residual values (not shown), which clearly show the increasing variation for the larger diamonds. The statistics at the end of the output are

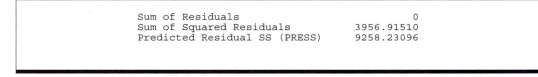

```
            Sum of Residuals                          0
            Sum of Squared Residuals         3956.91510
            Predicted Residual SS (PRESS)    9258.23096
```

The sum of residuals is zero, as it should be. The relative magnitude of the PRESS suggests that some observations are unduly influencing the estimated regression relationship.

3.4.1 Weighted Least Squares

One principle that may be used to reduce influence of highly variable observations is that of *weighted least squares*, which is a direct application of generalized least squares (Myers 1990, Section 7.1). The estimated parameters obtained by this method are those that minimize the weighted residual sum of squares:

$$\Sigma w_i \left(y_i - \beta_0 - \beta_1 x_1 - \dots \beta_m x_m\right)^2$$

where the w_i are a set of nonnegative weights assigned to the individual observations. Observations with small weights contribute less to the sum of squares and thus provide less influence to the estimation parameters, and vice versa for observations with larger weights. Thus, it is logical to assign small weights to observations whose large variances make them more unreliable, and likewise to assign larger weights to observations with smaller variances. It can, in fact, be shown that best linear unbiased estimates are obtained if the weights are inversely proportional to the variances of the individual errors.

The variances of the residuals, however, are not usually known. Multiple observations, or replicates, may be used to estimate this variance (Freund and Wilson 1998, Section 3.4), or if true replicates are not available, near neighbors (Montgomery and Peck 1982) may be used. However, these are all estimated variances, and both of the above references warn that the use of weights based on poor estimates of variances may be counterproductive.

Alternately, knowledge about the distribution of residuals may provide a basis for determining weights. It is well known that prices tend to vary by proportions. This implies that standard deviations are proportional to means; hence the variances are proportional to the squares of means. In sample data the means are not known, but estimated means based on an unweighted regression can be used for this purpose.

You can implement this method with the following SAS statements:

```
proc reg data=diamonds;
   model price = carats csq;
   output out = a p = pprice   lcl = lower ucl = upper;
data b; set a;
   w = 1/(pprice*pprice);
proc reg data = b; weight w;
   model price = carats csq / p;
output out=c p=wpprice lcl = wlower ucl = wupper;
run;
```

The first PROC REG produces the unweighted analysis that is given in Output 3.17. The OUTPUT statement creates the data set A in which the variable PPRICE is the price as predicted by the unweighted regression. (The prediction intervals will be used later.) The reciprocal of the square of PPRICE is required for the weighted regression; hence the variable PPRICE in data set A is used to produce the variable W=1/(PPRICE*PPRICE) in data set B. This variable is then used in the WEIGHT statement, which implements the weighted regression specified in the second PROC REG statement. The output data sets containing variables representing the prediction intervals for both models are requested for use in plots. The results of the weighted regression appear in Output 3.18.

Output 3.18
Weighted
Regression

```
                              The REG Procedure
                                Model: MODEL1
                          Dependent Variable: PRICE

                                 Weight: W

                            Analysis of Variance

                                    Sum of         Mean
      Source            DF         Squares       Square    F Value    Pr > F

      Model              2        14.24850      7.12425      33.40    <.0001
      Error             27         5.75948      0.21331
      Corrected Total   29        20.00799


              Root MSE              0.46186    R-Square     0.7121
              Dependent Mean        3.44598    Adj R-Sq     0.6908
              Coeff Var            13.40285


                           Parameter Estimates

                          Parameter      Standard
      Variable      DF     Estimate         Error    t Value    Pr > |t|

      Intercept      1      5.03651       5.14379       0.98      0.3362
      CARATS         1    -18.74381      13.22756      -1.42      0.1679
      CSQ            1     23.97856       7.96786       3.01      0.0056
```

A comparison of the results with those of the unweighted least squares regression (see Output 3.17) shows the effect of weighting. Since sums of squares are a function of the weights, they cannot be compared with those of the unweighted analysis. However, the R-Square statistics are comparable since they are ratios and the effect of the weights cancels. You can see that the R-Square values are somewhat larger for the weighted analysis; this is because the effect of the largest residuals has been reduced. Also the estimated equation has a smaller coefficient for $(CARATS)^2$. In other words, the curve has a somewhat smaller upward curvature, which is presumably due to the lesser influence of the very high-priced diamonds. The listing of predicted and residuals is not shown; the statistics at the bottom are

```
      Sum of Residuals                        0
      Sum of Squared Residuals          5.75948
      Predicted Residual SS (PRESS)     6.82899
```

The magnitude of the PRESS sum of squares is now not much larger than that of the residual sum of squares, showing the reduction of influence of the very large diamonds.

The major feature of the weighted analysis is that the prediction intervals are much narrower for the smaller diamonds than for the large diamonds. This can be seen by a plot comparing the prediction intervals for the unweighted and weighted analyses. You can produce these plots by adding the following SAS statements:

```
proc gplot data=c;
   symbol1 v=star c=black;
   symbol2 v=point l=1 i=spline c=blsck;
   plot pprice*carats=1 lower*carats=2 upper*carats=2
   /overlay;
   plot wpprice*carats=1 wlower*carats=2 wupper*carats=2 /
   overlay;
run;
```

The resulting plots in Output 3.19 clearly show the differences in the prediction intervals. Although all are possibly too wide to make the estimates very useful, the intervals based on weighted least squares do indicate better precision of the estimates for the smaller diamonds.

Output 3.19
Comparing
Prediction
Intervals

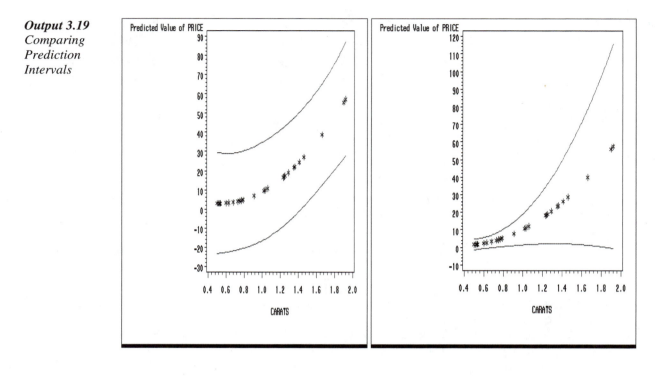

3.5 Correlated Errors

Another assumption regarding the distribution of the random errors is that the errors are independent. That is, the value of any one residual is independent of the value of any other. Most violations of this assumption occur in time series data, that is, data observed over a sequence of time periods. In such data the observed value at a given point in time, say, in period *t*, may be influenced by values observed in previous periods. For example, the weather today is highly influenced by what the weather was yesterday. Nonindependent errors may occur in other situations. For example, plants that are too close to each other may compete to the extent that smaller, stunted plants may surround larger plants.

In linear models, nonindependent errors are usually described as being correlated. If these correlations are known, generalized least squares (Rawlings, Pantula, and Dickey 1998, Section 12.5) can be used to provide correct estimates and other statistics. However, these correlations are in practice not known. In fact, these correlations cannot even be estimated since there are $(n)(n-1)/2$ such correlations, which greatly exceeds the number of observations (n) on which such estimates must be based.

This problem is alleviated somewhat by imposing a structure on the correlations. The most commonly used structure is the *autoregressive model*, which specifies that the error at time *t* is related to previous errors by a linear regression model. If the errors at time *t* are related only to errors at time (*t*−1), the model is called a first-order autoregressive model; if it is related to errors at time (t−1) and (*t*−2), it is called a second-order model; and so on. The most commonly used order is one; that is, the errors at time *t* are directly related only to the errors of the previous period.

It can be shown that if the existence of an autoregressive process is ignored, that is, if ordinary least squares is used for estimation and inference, the estimates of coefficients are unbiased, but estimates of the residual variance and standard errors of the partial coefficients are subject to an unknown, but usually downward, bias. It is therefore necessary to consider alternate analysis methodologies.[7] A rather simple approach is to base the regression on first differences (Freund and Wilson, 1998, Section 4.5). With this procedure, the variables are the period-to-period differences of all variables. In the SAS System, such variables can be created by the DIF function in the DATA step. This procedure does, however, change the model as it now relates period-to-period changes in the response to period-to-period changes in the independent variables. Therefore, results are not strictly comparable to those of the original model. This method is not illustrated here.

This section presents an example of one procedure for detecting the existence of a first-order autoregressive model, followed by an alternative analysis methodology provided by the AUTOREG procedure, described in the *SAS/ETS User's Guide*.

A consumption function is a model that attempts to estimate consumption of goods and services using other economic variables. Using quarterly data, consumption (CONS) is to be estimated by a regression on the independent variables:

CURR   currency in circulation

DDEP   amount of demand deposits

GNP   the gross national product

WAGES   the hourly wage rate

INCOME   national income.

All variables have been deflated by the consumer price index (January 1953=100), and all variables except WAGES have been converted to a per capita basis. The data set CONSUME consists of 76 observations from the first quarter of 1951 through the last quarter of 1969. The data appear in Output 3.20.

---

[7] The analysis of time series data comprises a major specialty within the discipline of statistics. The SAS/ETS documentation contains some of the most popular methods for analyzing time series data. Usage of the SAS System for these methods is presented in Brocklebank and Dickey (1986).

Output 3.20
Consumption
Data

YR	QTR	CURR	DDEP	GNP	WAGES	CONS
51	1	181.3	664.5	2296.9	1.713	1256.1
51	2	181.1	665.0	2332.2	1.739	1247.7
51	3	181.7	664.0	2353.4	1.736	1254.5
51	4	180.7	666.3	2340.8	1.746	1256.9
52	1	183.2	677.2	2365.2	1.779	1265.9
52	2	182.3	674.3	2333.1	1.764	1276.3
52	3	182.3	673.3	2350.5	1.796	1287.5
52	4	184.2	676.4	2422.1	1.828	1306.8
53	1	186.3	678.2	2467.4	1.868	1321.1
53	2	185.6	674.6	2461.8	1.865	1317.7
53	3	184.2	668.0	2424.2	1.874	1309.5
53	4	183.9	668.1	2386.8	1.891	1307.9
54	1	182.6	669.0	2377.3	1.891	1320.1
54	2	180.7	666.6	2360.1	1.898	1322.8
54	3	179.8	674.7	2384.2	1.904	1338.2
54	4	178.9	683.1	2438.5	1.942	1355.1
55	1	178.8	689.3	2511.4	1.953	1367.5
55	2	178.7	690.9	2553.9	1.974	1379.2
55	3	177.7	687.9	2582.7	2.009	1382.2
55	4	177.7	685.9	2613.4	2.032	1408.3
56	1	177.6	685.6	2614.0	2.053	1422.2
56	2	174.6	676.0	2605.2	2.059	1414.0
56	3	173.2	667.9	2601.1	2.075	1416.8
56	4	171.5	661.7	2621.6	2.100	1419.8
57	1	170.2	655.4	2636.7	2.095	1428.5
57	2	167.6	646.0	2614.3	2.082	1416.8
57	3	166.2	637.9	2621.3	2.087	1429.0
57	4	164.8	628.3	2570.7	2.099	1427.7
58	1	161.3	616.5	2485.8	2.070	1410.8
58	2	160.8	621.0	2490.4	2.083	1423.9
58	3	160.6	625.6	2553.1	2.103	1439.5
58	4	160.5	631.7	2614.7	2.153	1447.0
59	1	161.0	635.4	2658.2	2.173	1464.8
59	2	160.3	633.0	2701.0	2.177	1469.0
59	3	159.4	629.8	2659.8	2.147	1478.3
59	4	157.6	619.7	2675.6	2.190	1482.1
60	1	157.5	609.9	2732.0	2.207	1494.2
60	2	155.8	598.6	2712.0	2.192	1505.1
60	3	154.8	597.5	2692.2	2.197	1498.3
60	4	153.3	592.0	2660.1	2.204	1500.5
61	1	152.2	593.4	2651.5	2.204	1508.5
61	2	151.4	595.8	2698.1	2.231	1513.8
61	3	151.0	592.9	2719.2	2.218	1514.2
61	4	152.0	597.1	2779.8	2.268	1532.9
62	1	152.2	594.6	2807.9	2.267	1536.7
62	2	152.8	592.0	2838.8	2.270	1544.7
62	3	152.0	583.0	2841.6	2.253	1546.7
62	4	152.9	585.5	2877.3	2.297	1569.4
63	1	154.4	587.5	2884.3	2.298	1577.5
63	2	155.2	587.5	2896.6	2.308	1580.7
63	3	156.3	587.3	2922.5	2.306	1591.2
63	4	157.4	587.7	2952.3	2.333	1588.3
64	1	159.2	587.8	2998.1	2.331	1619.2
64	2	160.7	587.7	3030.1	2.343	1631.3
64	3	161.9	592.0	3060.2	2.362	1659.2
64	4	162.1	593.9	3067.4	2.371	1663.7
65	1	163.8	593.6	3137.2	2.376	1677.7
65	2	163.1	590.1	3155.6	2.371	1698.3
65	3	165.1	594.0	3209.3	2.387	1720.0
65	4	166.2	597.5	3261.1	2.396	1742.4
66	1	167.0	600.2	3303.4	2.393	1758.9
66	2	168.0	599.8	3317.4	2.400	1771.0
66	3	167.9	587.7	3326.2	2.401	1775.9
66	4	168.3	581.5	3357.4	2.415	1773.6
67	1	169.6	582.9	3358.6	2.426	1800.9
67	2	169.5	586.0	3359.4	2.431	1803.9
67	3	169.1	595.4	3386.8	2.434	1808.1
67	4	169.5	596.1	3450.5	2.462	1807.0
68	1	170.3	595.2	3486.3	2.477	1841.4
68	2	171.2	597.1	3533.6	2.481	1845.6
68	3	172.1	602.0	3557.5	2.488	1863.6
68	4	172.4	601.2	3570.0	2.514	1857.6
69	1	173.4	600.4	3572.8	2.492	1862.1
69	2	173.0	595.3	3570.3	2.484	1861.6
69	3	171.8	584.1	3583.1	2.506	1862.2
69	4	171.7	588.5	3553.2	2.506	1862.1

Assuming that the data are in sequence, obtain the usual regression as follows:

```
proc reg data = consume;
   model cons = curr ddep gnp wages income / dw;
run;
```

The MODEL statement option DW requests the computation of the Durbin-Watson *d* statistic for testing for the existence of a first-order autoregressive process. The results of the procedure appear in Output 3.21.

Output 3.21
Regression
for
Estimating
Consumption
Function

```
                          The REG Procedure
                            Model: MODEL1
                       Dependent Variable: CONS

                         Analysis of Variance

                              Sum of        Mean
Source               DF      Squares       Square    F Value    Pr > F

Model                 5      2702414       540483    3366.92    <.0001
Error                70        11237    160.52719
Corrected Total      75      2713651


           Root MSE             12.66993    R-Square     0.9959
           Dependent Mean     1532.12503    Adj R-Sq     0.9956
           Coeff Var             0.82695


                         Parameter Estimates

                      Parameter      Standard
Variable     DF        Estimate         Error    t Value    Pr > |t|

Intercept     1       160.38765      95.79101       1.67      0.0985
CURR          1         1.48289       0.60087       2.47      0.0160
DDEP          1        -0.57537       0.14215      -4.05      0.0001
GNP           1         0.11587       0.11466       1.01      0.3157
WAGES         1       323.56479      55.09253       5.87      <.0001
INCOME        1         0.19256       0.11831       1.63      0.1081

                  Durbin-Watson D              0.949
                  Number of Observations          76
                  1st Order Autocorrelation    0.518
```

The model obviously fits the data quite well. The most important coefficients are those for demand deposits and wages. However, the negative coefficient for demand deposits is inconsistent with economic theory.

The last two items on the output concern autocorrelation. The Durbin-Watson *d* statistic is a test for existence of a first-order autoregressive process. According to a table of this statistic (Freund and Wilson 1998, Appendix Table A.5), a value of less than 1.60 indicates existence of a positive first-order autocorrelation at a significance level of less than 0.01. (As you will see below, *p*-values for the Durbin-Watson statistic are available in PROC AUTOREG.) The second item is the actual sample correlation of adjacent residuals: the value of 0.518 shows a moderate degree of association between adjacent residuals. This first-order autocorrelation is also known as the *lag-1* autocorrelation, as it is the correlation of observations with those that lag behind one period.

A plot of residuals against time may also be useful in detecting autocorrelation. To do this, create a new data set called PLOT as follows:

```
data plot; set a;
n = _n_;
run;
```

This DATA step uses the automatic observation variable _N_ to create the sequential period indicator, *n*. Because the relationship between adjoining data points is of interest, time series plots are more useful if done by SAS/GRAPH software as follows:[8]

```
symbol1 v=point i=join c=black;
proc gplot data=plot;
   plot rc*n =1 / vref=0;
run;
```

The plot is shown in Output 3.22.

Output 3.22
Residual
Plot

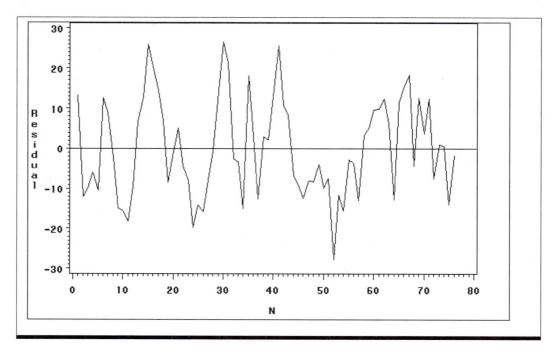

The plot appears to show irregular cycles, that is, a larger than expected number of sequences of either positive or negative residuals. This is due to the fact that under a first-order autoregressive process with a positive autocorrelation, a positive residual is more likely to be followed by another positive residual and likewise for negative residuals. Of course, apparent cycles in the residuals may also be caused by true cyclical effects, such as seasons.

One popular alternative analysis procedure for data of this type is available in PROC AUTOREG, a procedure in SAS/ETS software. This procedure uses the residuals from an ordinary least squares analysis to estimate the set of autoregressive parameters for the order of autoregressive model specified by the user. These coefficients are then used to perform the appropriate generalized least squares analysis, which is implemented by performing a transformation of the variables in the model (Fuller 1978, Section 2.5).

---

[8] If SAS/GRAPH software is not available, a plot of this type is more useful if each vertical position of the plot represents one period, which you can usually do with the HAXIS option.

Although the primary focus is on the first-order autoregressive model, it is still useful to investigate the possibility of higher orders. To do this, you can specify

```
proc autoreg data=consume;
   model cons = curr ddep gnp wages income/ dw=4 dwprob
   nlag=4;
run;
```

The statements are in the same form as PROC REG. The MODEL statement options DW=4 and DWPROB request the Durbin-Watson statistics and *p*-values for lags 1 through 4 and NLAG=4 specifies that a fourth-order autoregressive model is to be used. This model is used to allow detection of seasonal cycles. Other MODEL and PROC options are discussed below. Output 3.23 shows the output from PROC AUTOREG.

Output 3.23
PROC
AUTOREG
Output for
Consumption
Function

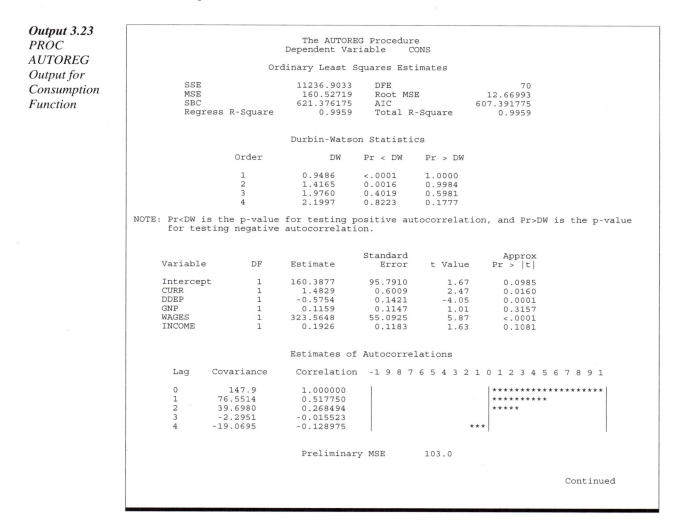

```
                          The AUTOREG Procedure
                     Dependent Variable     CONS

                     Ordinary Least Squares Estimates

        SSE                  11236.9033    DFE                       70
        MSE                   160.52719    Root MSE            12.66993
        SBC                  621.376175    AIC                607.391775
        Regress R-Square         0.9959    Total R-Square        0.9959


                        Durbin-Watson Statistics

               Order          DW      Pr < DW      Pr > DW

                 1        0.9486       <.0001       1.0000
                 2        1.4165       0.0016       0.9984
                 3        1.9760       0.4019       0.5981
                 4        2.1997       0.8223       0.1777

NOTE: Pr<DW is the p-value for testing positive autocorrelation, and Pr>DW is the p-value
      for testing negative autocorrelation.


                                    Standard                Approx
       Variable      DF    Estimate    Error    t Value    Pr > |t|

       Intercept      1    160.3877   95.7910      1.67      0.0985
       CURR           1      1.4829    0.6009      2.47      0.0160
       DDEP           1     -0.5754    0.1421     -4.05      0.0001
       GNP            1      0.1159    0.1147      1.01      0.3157
       WAGES          1    323.5648   55.0925      5.87      <.0001
       INCOME         1      0.1926    0.1183      1.63      0.1081


                        Estimates of Autocorrelations

       Lag    Covariance    Correlation  -1 9 8 7 6 5 4 3 2 1 0 1 2 3 4 5 6 7 8 9 1

        0        147.9        1.000000     |                   |********************|
        1        76.5514      0.517750     |                   |**********          |
        2        39.6980      0.268494     |                   |*****               |
        3        -2.2951     -0.015523     |                   |                    |
        4       -19.0695     -0.128975     |               ***|                     |


                        Preliminary MSE        103.0


                                                            Continued
```

Output 3.23 (Continued) PROC AUTOREG Output for Consumption Function

```
                    Estimates of Autoregressive Parameters

                                      Standard
            Lag      Coefficient         Error      t Value

             1        -0.505172        0.122880       -4.11
             2        -0.116444        0.135901       -0.86
             3         0.181086        0.135901        1.33
             4         0.058640        0.122880        0.48
                      The AUTOREG Procedure

                     Yule-Walker Estimates

      SSE              7618.97051    DFE                      66
      MSE               115.43895    Root MSE           10.74425
      SBC              609.631443    AIC               586.32411
      Regress R-Square     0.9926    Total R-Square       0.9972


                    Durbin-Watson Statistics

            Order              DW      Pr < DW      Pr > DW

             1             1.9009       0.1962       0.8038
             2             1.8498       0.1954       0.8046
             3             1.9308       0.3493       0.6507
             4             1.9999       0.5241       0.4759
NOTE: Pr<DW is the p-value for testing positive autocorrelation, and Pr>DW is the p-value for
      testing negative autocorrelation.


                                      Standard                   Approx
        Variable        DF     Estimate      Error    t Value    Pr > |t|

        Intercept        1     262.0764    123.9299      2.11      0.0382
        CURR             1       0.9220      0.8225      1.12      0.2664
        DDEP             1      -0.5179      0.2044     -2.53      0.0137
        GNP              1       0.1565      0.1433      1.09      0.2789
        WAGES            1     261.0441     70.7811      3.69      0.0005
        INCOME           1       0.1831      0.1476      1.24      0.2192
```

The upper portion of the output reproduces the coefficients estimated by ordinary least squares (often denoted by OLS), which are the same as appear in Output 3.19. The Durbin-Watson statistic and *p* values strongly suggest a positive first-order autocorrelation and moderately suggest a positive second-order autocorrelation. The residuals from this regression are used to compute the autocorrelations for the lags specified by the NLAG option that are printed under the heading "Estimates of Autocorrelations." The correlation for lag 1 is the same as appears in Output 3.19.

The plot of the autocorrelations provides for a quick check on possible patterns of the various autocorrelations. For example, if there is only a first-order autocorrelation with a correlation of ρ, then it can be shown that the second-order (lag 2) autocorrelation is ρ^2, the third-order correlation is ρ^3, and so forth. For a positive first-order process, then, the plot would show the correlations tending rapidly to zero with increasing lag as they do here.

These correlations are used to estimate the autoregressive parameters, that is, the regression coefficients for the model that relates the residuals of period *t* to those of the previous periods. These coefficients appear next in the output, together with their standard deviations (standard errors) and *t* ratios for testing that they are zero. These statistics confirm that there appears to be only a first-order autoregressive process, and they suggest that the model should be reestimated using the option NLAG=1. This alternative is not illustrated here.

The autoregressive parameter estimates are used to perform the transformations of the variables required to obtain the generalized least squares estimates. The transformed variables are not printed but are available in an output data set. The resulting estimates comprise the final portion of the output. A comparison of these estimates and their standard errors with those of the ordinary least squares method shows that they are not very different. Since the upward bias in variance estimates for a first-order autoregressive model is generally accepted to be in the neighborhood of $1/(1-\rho^2)$, these results are not surprising. In other words, if the first-order autocorrelation is 0.5, variance estimates using ordinary least squares should be approximately 33% larger than they actually are. The Durbin-Watson statistics for this regression now indicates the autocorrelations have essentially vanished.

It is interesting to compare the PROC REG and PROC AUTOREG predictions. The following steps accomplish this:

```
proc reg data = consume noprint;
    model cons = curr ddep gnp wages income ;
    output out=a r=rcons;
proc autoreg data = consume noprint;
    cons = curr ddep gnp wages income / nlag = 4;
    output out=b r = racons;
    data c; merge a b;
    symbol1 v=point i=join l=1 c=black;
    symbol2 v=point I=join l=33 c=black;
proc gplot data=c;
    plot rcons*n=1 racons*n=2 / overlay;
run;
```

The resulting plot is hown in Output 3.24.

Output 3.24
Comparing
Residuals

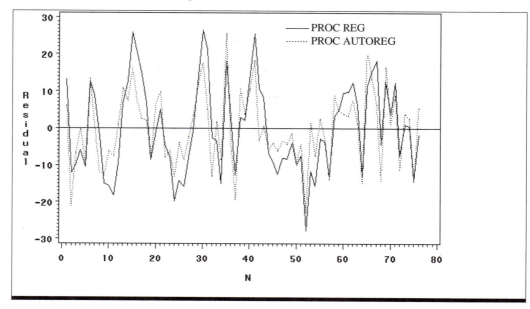

It does appear that the magnitudes of the residuals from the ordinary least squares regression have been reduced somewhat by the use of PROC AUTOREG, but the major difference is that the lengths of the sequences of residuals above and below the mean have been reduced. In fact, the first-order autocorrelation among these residuals is almost exactly zero.

PROC AUTOREG has a number of PROC and MODEL statement options relating to output of various intermediate results. MODEL statement options may be used to restrict the number of autoregressive parameters to be estimated and also to perform, if desired, a backward elimination of autoregressive parameters. The procedure can also be requested to produce an output data set containing the parameter estimates and the transformed variables. These transformed variables may be used as input for repeated implementations of PROC AUTOREG, thus providing an iterative procedure for finding better estimates.

3.6 Summary

This chapter is titled "Observations" because it is concerned with aspects of observations that can cause a regression analysis to be of questionable value. Most of the focus has been on problems with the random error, but it was shown that problems may exist with the independent variables as well.

The problems are of four major types that may occur singly or in combination:

❑   outliers

❑   specification error

❑   heterogeneous variances

❑   nonindependent errors.

The most difficult of these is the problem of outliers and influential observations, a point that is illustrated by the large number of statistics that have been developed to detect them. Unfortunately, none of these is uniformly superior because outliers and influential observations can exist in so many ways. Finally, if you have found an outlier or influential observation (or several) there is no guide as to what action to take.

The other difficulties are somewhat easier to diagnose and remedy although, even for these, a degree of subjectivity is often needed to provide a reasonable analysis.

This chapter has demonstrated how the SAS System can be used to detect violations of assumptions about the error term in the regression model and subsequently to employ some remedial methods. Not only can the violations of assumptions take many forms, but also the nature of violations is not always distinct. For example, apparent outliers may be the result of a nonnormal distribution of errors. Therefore, it is important not to use these methods blindly by taking a scattershot approach of doing everything. Instead, methods must be chosen carefully, often on the basis of prior information. Such information may include knowledge of distributions that are known to exist with certain types of data or outliers resulting from sloppiness that occur with some regularity from some data sources. As usual, the power of even the most sophisticated computer software is no substitute for the application of human intelligence.

Chapter 4 Multicollinearity: Detection and Remedial Measures

4.1 Introduction

The validity of inferences resulting from a regression analysis assumes the use of a model with a specified set of independent variables. But in many cases you do not know specifically what variables should be included in a model. Hence, you may propose an initial model, often containing a large number of independent variables, and use a statistical analysis in hopes of revealing the correct model. This approach has two major problems:

❑ The model and associated hypotheses are generated by the data, thus invalidating significance levels (*p* values). This problem is related to the control of Type I errors in multiple comparisons, and although the problem has been partially solved for that application, it has not been solved for variable selection in regression. Therefore, although *p* values are provided by computer outputs for models obtained by such methods, they cannot be taken literally.

❑ The inclusion of a large number of variables in a regression model often results in *multicollinearity*, which is defined as a high degree of correlation among several independent variables.[1] This occurs when too many variables have been put into the model and a number of these variables measure similar phenomena.

---

[1] Correlations among independent variables also occur in polynomial models (see Chapter 5) where, for example, there may be a high positive correlation between x and x^2. Remedial measures for this type of multicollinearity are not considered in this chapter.

The existence of multicollinearity is not a violation of the assumptions underlying the use of regression analysis. In other words, the existence of multicollinearity does not affect the estimation of the dependent variable. That is, the resulting $\hat{y}$ values are the best linear unbiased estimates of the conditional means of the population. Depending on the purpose of the regression analysis, multicollinearity may, however, limit the usefulness of the results as follows:

❑   The existence of multicollinearity tends to inflate the variances of predicted values, that is, predictions of the response variable for sets of *x* values. This inflation may be especially severe when the values of the independent variables are not in the sample.

❑   The existence of multicollinearity tends to inflate the variances of the parameter estimates. A partial regression coefficient measures the effect of the corresponding independent variable, holding constant all other variables. Now if there is a high degree of correlation among the independent variables, this condition does not occur in the data. In other words, when multicollinearity exists, partial coefficients are trying to estimate something that does not occur in the data. Inadequate data tend to generate unstable coefficients that may have incorrect signs or magnitudes (Freund and Wilson 1998, Section 5.2). This is an especially troublesome result when it is important to ascertain the structure of the relationship of the response to the various independent variables.

Because the use and interpretation of regression coefficients are very important for many regression analyses, you should ascertain if multicollinearity exists. Then, if multicollinearity is deemed to exist, it is important to ascertain the nature of the multicollinearity, that is, the nature of the linear relationships among the independent variables. The final step is to alleviate the effects of multicollinearity.

Section 4.2, "Detecting Multicollinearity," presents some methods available in PROC REG for the detection of multicollinearity. You will see that it is not difficult to detect, but once detected, the nature of the multicollinearity may be more difficult to diagnose.

Alleviating the effects of multicollinearity is even more difficult, and the methods used depend on the ultimate purpose of the regression analysis. In fact, there is no universally optimum strategy for this task, and, furthermore, results obtained from any strategy are often of questionable validity and usefulness.

When it is important to study the nature of the regression relationship, it may be useful to redefine the variables in the model. Such redefinition may be simply based on knowledge of the variables or on a multivariate analysis of the independent variables. Variable redefinition procedures are presented in Section 4.3.1, "Redefining Variables."

The most obvious and therefore most frequently used strategy is to implement a model with fewer independent variables. Since there is often no a priori criterion for the selection of variables, an automated, data-driven search procedure is most frequently used. Implementation of such variable-selection procedures is often called *model building*. However, since these methods use the data to select the model, they are more appropriately called *data dredging*. This approach has two major drawbacks:

❑   It is not appropriate if you are trying to determine the structure of the regression relationship.

❑   The results of the variable selection procedures often do not provide a clear choice of an optimum model.

Procedures for variable selection using the SAS System are presented in Section 4.4, "Variable Selection."

Another method that may be useful for studying the structure of the regression relationship is biased estimation. Although least squares is known to provide unbiased estimates of the parameters of a linear model, there are situations in which biased estimates may have smaller variances and thus provide useful estimates. Multicollinearity is a major contributor to large variances of estimates of regression coefficients. Hence, a biasing of estimates that reduces the effect of multicollinearity may provide estimates with smaller mean-squared error. Two popular biased estimators are presented in Section 4.5, "Biased Estimation."

4.2 Detecting Multicollinearity

The example used for detecting and alleviating the effects of multicollinearity is the BOQ data used for outlier detection in Chapter 3, omitting the impossible observation 25.

Three sets of statistics that may be useful for ascertaining the degree and nature of multicollinearity are available as MODEL statement options in PROC REG. These statistics are

❑   the relative degree of significance of the model and the individual parameter estimates

❑   the variance inflation factors, often referred to as VIF

❑   an analysis of the structure of the $\mathbf{X'X}$ matrix.

PROC REG provides these statistics when you submit the following statements:

```
proc reg data=boq;
   model manh = occup checkin hours common wings cap rooms /
      vif collinoint;
run;
```

The VIF option requests the calculation of the variance inflation factors that are appended to the listing of parameter estimates, as seen in Output 4.1 and as discussed in Section 4.2.1, "Variance Inflation Factors." The COLLINOINT (or COLLIN) option, shown in Output 4.2, provides for the analysis of the $\mathbf{X'X}$ structure. These statistics are discussed in Section 4.2.2, "Analysis of Structure: Principal Components, Eigenvalues, and Eigenvectors."

Output 4.1
Regression for
BOQ Data,
Omitting
Observation 25

```
                                    The REG Procedure
                                      Model: MODEL1
                                Dependent Variable: MANH

                                    Analysis of Variance

                                        Sum of         Mean
     Source                   DF       Squares       Square     F Value    Pr > F

     Model                     7      87497673     12499668      154.32    <.0001
     Error                    16       1295987        80999
     Corrected Total          23      88793659


                 Root MSE               284.60353    R-Square      0.9854
                 Dependent Mean        2050.00708    Adj R-Sq      0.9790
                 Coeff Var               13.88305


                                    Parameter Estimates

                          Parameter      Standard                            Variance
     Variable     DF       Estimate         Error    t Value    Pr > |t|    Inflation

     Intercept     1      171.47336     148.86168       1.15      0.2663            0
     OCCUP         1       21.04562       4.28905       4.91      0.0002     43.63222
     CHECKIN       1        1.42632       0.33071       4.31      0.0005      4.54154
     HOURS         1       -0.08927       1.16353      -0.08      0.9398      1.36076
     COMMON        1        7.65033       8.43835       0.91      0.3781      4.06083
     WINGS         1       -5.30231       9.45276      -0.56      0.5826      3.79996
     CAP           1       -4.07475       3.30195      -1.23      0.2350     56.60333
     ROOMS         1        0.33191       6.81399       0.05      0.9618    178.70159
```

A comparison of the relative degrees of statistical significance of the model with those of the partial regression coefficients reveals multicollinearity. The overall model is highly significant with an *F* value of 154.3 and *p* value much smaller than 0.0001. The smallest *p* value for a partial regression coefficient is 0.0002, which is certainly not large but is definitely larger than that for the overall model. This type of result is a natural consequence of multicollinearity: the overall model may fit the data quite well, but because several independent variables are measuring similar phenomena, it is difficult to determine which of the individual variables contribute significantly to the regression relationship.

4.2.1 Variance Inflation Factors

The variance inflation factors are useful in determining which variables may be involved in the multicollinearities. For the *i*th independent variable, the variance inflation factor is defined as $1/(1 - R_i^2)$, where R_i^2 is the coefficient of determination for the regression of the *i*th independent variable on all other independent variables. It can be shown (Freund and Wilson 1998, Section 5.3) that the variance of the estimate of the corresponding regression coefficient is larger by that factor than it would be if there were no multicollinearity. In other words, the VIF statistics show how multicollinearity has increased the instability of the coefficient estimates.

There are no formal criteria for determining the magnitude of variance inflation factors that cause poorly estimated coefficients. Some authorities suggest that values exceeding 10 may be cause for concern, but this value is arbitrary. Actually, for models with low coefficients of determination for the regression, estimates of coefficients that exhibit relatively small variance inflation factors may still be unstable, and vice versa. In Output 4.1 the regression R^2 is a rather high, 0.9854. Since $1/(1 - R^2) = 68.5$, any variables associated with VIF values exceeding 68.5 are more closely related to the other independent variables than they are to the dependent variable. Only the coefficient ROOMS has a VIF value larger than 68.5, and it is certainly not statistically significant. However, the variable OCCUP has a VIF value larger than 10 and a very small *p* value. You may conclude, therefore, that multicollinearity exists that decreases the reliability of the coefficient estimates. The overall regression is so strong, however, that some coefficients may still be meaningful. On the other hand, the variable CAP has a rather large VIF and a large *p* value, so this variable may have been useful had it not been involved in multicollinearity.

4.2.2 Analysis of Structure: Principal Components, Eigenvalues, and Eigenvectors

Both the analysis of the structure of multicollinearity and attempts to alleviate the effects of multicollinearity (Section 4.3.2, "Multivariate Structure: Principal Component Regression") make use of a multivariate statistical technique called *principal components*.

Multivariate analysis is the study of variation involving multiple response variables. *Principal component analysis* is a multivariate analysis technique that attempts to describe interrelationships among a set of variables. Starting with a set of observed values on a set of *m* variables, this method uses linear transformations to create a new set of variables, called the principal components, which are defined as

$$Z = X\Gamma$$

where

> Z is the matrix of principal component variables
>
> X is the matrix of observed variables
>
> Γ is the matrix of coefficients that relate Z to X.

The principal component transformation has the following properties:

❑   The principal component variables, or simply components, are jointly uncorrelated.

❑   The first principal component has the largest variance of any linear function of the original variables (subject to a scale constraint). The second component has the second largest variance, and so on.

Because the principal component variables are uncorrelated, they exhibit absolutely no multicollinearity but are not guaranteed to provide useful interpretations.

Principal components are obtained by computing the eigenvalues and eigenvectors of the correlation or covariance matrix. In most applications the correlation matrix is used so that the scales of measurement of the original variables do not affect the components.

The eigenvalues are the variances of the components. If the correlation matrix has been used, the variance of each input variable is one; hence the sum of the variances is equal to the number of variables. Because of the scale constraint of the principal component transformation, the sum of variances (eigenvalues) of the component variables is also equal to the number of variables, but the variances are not equal.

The eigenvectors, which are the columns of Γ, are the coefficients of the linear equations that relate the component variables to the original variables.

A set of eigenvalues of relatively equal magnitudes indicates that there is little multicollinearity, while a wide variation in magnitudes indicates severe multicollinearity. In fact, the number of large (usually greater than unity) eigenvalues may be taken as an indicator of the true number of variables (sometimes called factors) needed to describe the behavior of the full set of variables. In other words, a small number of large eigenvalues indicates that a small number of component variables describes most of the variability of the originally observed variables. Because of the scale constraint, a number of large eigenvalues implies that there will be some small eigenvalues, and such values imply that some component variables have small variances. In fact, zero-valued eigenvalues indicate that exact collinearity (Section 2.9) exists among the x variables; hence the existence of very small eigenvalues implies severe multicollinearity. For more complete descriptions of principal components, see Johnson and Wichern (1982) and Morrison (1976).

Principal component variables with very small variances are of interest in identifying sources of multicollinearity, as shown in this section. Principal component variables with large variances are used in attempts to provide better interpretation of the regression, as shown in Section 4.3.2.

The *analysis of structure* of relationships among a set of variables is afforded by the study of the eigenvalues and eigenvectors of $\mathbf{X'X}$. This analysis can be performed in two ways:

❑   by using the raw (not centered) variables, including the dummy variable used to estimate the intercept, and scaling the variables such that the $\mathbf{X'X}$ matrix has ones on the diagonal. This method is implemented by the COLLIN option in the MODEL statement.

❑ by using scaled and centered variables and excluding the dummy variable. In this case, **X′X** is the correlation matrix. This method is implemented by the COLLINOINT option in the MODEL statement.

The first method suggests that the dummy variable used to estimate the intercept is simply one of the variables that can be involved in multicollinearity. This is not an appropriate conclusion for many situations. The intercept is an estimate of the response at the origin, that is, where all independent variables are zero. It is chosen for mathematical convenience rather than to provide a useful parameter. In fact, for most applications the intercept represents an extrapolation far beyond the reach of the data. For this reason the inclusion of the intercept in the study of multicollinearity can be useful only if the intercept has some physical interpretation and is within reach of the actual data space.

Centering the variables places the intercept at the means of all the variables. Therefore, if the variables have been centered, the intercept has no effect on the multicollinearity of the other variables (Belsley, Kuh, and Welsch 1980). Centering is also consistent with the computation of the variance inflation factors which are based on first centering the variables. Since the origin cannot exist in the BOQ example, the COLLINOINT option is used here.

The COLLINOINT option provides the eigenvectors and variance proportions associated with the eigenvalues. Output 4.2 shows the portion of the output from the COLLINOINT option.

Output 4.2
COLLINOINT
Option

```
              Collinearity Diagnostics(intercept adjusted)

                        Condition ------------Proportion of Variation-------------
   Number  Eigenvalue   Index      OCCUP       CHECKIN       HOURS        COMMON
     1      5.04674    1.00000   0.00082851    0.00649     0.00970      0.00505
     2      0.72303    2.64197   0.00000824    0.00948     0.70454      0.04625
     3      0.69939    2.68624   0.00164       0.05107     0.23867      0.13255
     4      0.31781    3.98494   0.00262       0.05098     0.02636      0.28025
     5      0.15871    5.63909   0.00165       0.77761     0.00643      0.00490
     6      0.05049    9.99779   0.27170       0.10408     0.00341      0.05860
     7      0.00384   36.25398   0.72155    0.00028999     0.01088      0.47240

              Collinearity Diagnostics(intercept adjusted)

                         --------Proportion of Variation---------
             Number       WINGS          CAP           ROOMS
               1         0.00686      0.00061879     0.00021117
               2         0.04961      3.276095E-7    0.00002561
                         The REG Procedure
                          Model: MODEL1
                       Dependent Variable: MANH

              Collinearity Diagnostics(intercept adjusted)

                         --------Proportion of Variation---------
             Number       WINGS          CAP           ROOMS
               3         0.03443       0.00167       0.00021312
               4         0.41437      0.00085179     0.00002193
               5         0.11620       0.02905        0.00326
               6         0.36479       0.10251       0.00026446
               7         0.01376       0.86530        0.99600
```

The first column of this output consists of the eigenvalues of the correlation matrix of the set of independent variables. The eigenvalues are arranged from largest to smallest. The severity of multicollinearity is revealed by the relative magnitudes of these eigenvalues. Large variability among the eigenvalues indicates a greater degree of multicollinearity. Two features of these eigenvalues are of interest:

❑ Eigenvalues of zero indicate linear dependencies or exact collinearities. Therefore, very small eigenvalues indicate near-linear dependencies or high degrees of multicollinearity. There are two eigenvectors that may be considered very small in this set, implying the possibility of two sets of very strong relationships among the variables.

❑ The last element in the column labeled Condition Index, sometimes called the condition number, is the square root of the ratio of the largest to smallest eigenvalue. This value provides a single statistic for indicating the severity of multicollinearity. Criteria for a condition number to signify serious multicollinearity are arbitrary, with the value 30 often quoted. Myers (1990, Chapter 7) mentions that the square of the condition index in excess of 1000 indicates serious multicollinearity. [2] The condition number of 36.25 for this example indicates considerable multicollinearity.

The other elements of the Condition Index column are the square roots of the ratio of the largest to each of the other eigenvalues. The number of large values in this column also indicates near-linear dependencies among the variables.

The remainder of the output, labeled Proportion of Variation, provides information to identify which variables are involved in the near-linear dependencies. These numbers tell you how much of the variance of the parameter estimate (given by the column heading) is associated with each eigenvalue. For example, looking at the seventh eigenvalue, you see the value 0.721 in the OCCUP column. This shows that 72% of the variance of the OCCUP parameter estimate is associated with eigenvalue number 7. Similarly, 27% is associated with eigenvalue number 6, and so on. In other words, relatively large proportion values in any row corresponding to a large condition number (small eigenvalue) may specify variables involved in that near-linear dependency.

One approach to the analysis of these variance proportions (Belsley, Kuh, and Welsch 1980) is to identify eigenvalues having condition numbers greater than 30. Then, variables with variance proportions larger than 0.5 for each of these eigenvalues are considered to be involved in the near-linear dependency that gives rise to these large condition numbers.

In this example, the condition number is 36.25 (>30) in the row for eigenvalue 7 (0.00384). The variance proportions are clearly large for the variables OCCUP, CAP, and ROOMS, implying a near-linear dependency among these variables, a result that was to be expected. These variables also have the largest variance inflation factors. The next-largest condition number is 9.997 (<30). Correspondingly, the variance proportions for row 6 do not have any clearly large values. Therefore, this analysis indicates that there exists only one set of strongly related variables.

4.3 Model Restructuring

In many cases redefining the model may alleviate the effects of multicollinearity. For example, in the analysis of economic time series data, variables such as GNP, population, production of steel, and so on, tend to be highly correlated because all are affected by inflation and the total size of the economy. Deflating such variables by a price index and population reduces multicollinearity. Model redefinition is illustrated with the BOQ data in Section 4.3.1.

---

[2] The COLLIN option produces larger condition numbers, especially if the origin is outside the region of the data.

In situations where a basis for model redefinition is not obvious, it may be useful to implement multivariate techniques to study the structure of multicollinearity and consequently to use the results of such a study to provide a better understanding of the regression relationships. One such multivariate method is principal component analysis, which generates a set of artificial uncorrelated variables that can then be used in a regression model. A principal component regression using the SAS System is presented in Section 4.3.2, and a biased method based on principal components is presented in Section 4.5.1, "Incomplete Principal Component Regression."

4.3.1 Redefining Variables

The various multicollinearity statistics in the BOQ data demonstrate that many of the variables are related to the size of the establishment. The regression, then, primarily illustrates the obvious relationship that larger establishments require more manpower. It is, however, more interesting to ascertain what other characteristics of Bachelor Officers Quarters require more or less manpower. You may be able to answer this question by redefining the variables in the model to measure per-room characteristics; that is, estimate the relationship of the man-hour requirement per room to the occupancy rate per room, and so on. Create the following variables in the DATA step:

```
data rel; set boq;
    relocc = occup / rooms;
    relcheck = checkin / rooms;
    relcom = common / rooms;
    relwings = wings / rooms;
    relcap = cap / rooms;
    relman = manh / rooms;
proc print;
    var relocc--relcap rooms hours relman;
run;
```

Output 4.3 shows the resulting variables along with the HOURS variable, which is not redefined, and also the ROOMS variable.

Output 4.3
BOQ Data
with
Redefined
Variables

Obs	RELOCC	RELCHECK	RELCOM	RELWINGS	RELCAP	ROOMS	HOURS	RELMAN
1	0.33333	0.6667	0.21000	0.16667	1.00000	6	4.0	30.0383
2	0.60000	0.3160	0.25000	0.20000	1.00000	5	40.0	36.5220
3	0.21200	0.0668	0.31160	0.12000	1.00000	25	42.5	7.9968
4	0.87500	0.2963	0.12500	0.12500	0.87500	8	168.0	35.5688
5	0.86842	0.4342	0.05895	0.10526	1.00000	19	168.0	14.0726
6	1.27692	1.8292	0.07692	0.07692	1.00000	13	40.0	12.6446
7	0.71917	0.0833	0.00000	0.08333	1.00000	36	40.0	27.7525
8	0.67915	0.8528	0.11745	0.12766	1.00000	47	168.0	19.8264
9	0.51468	0.6605	0.35545	0.12987	1.00000	77	40.0	12.2625
10	0.92542	3.3281	0.01250	0.37500	1.00000	48	168.0	22.9842
11	0.82697	3.1376	0.11773	0.09091	1.00000	66	168.0	21.0276
12	1.53054	10.0924	0.16297	0.10811	0.97297	37	168.0	40.2568
13	0.48469	1.8776	0.15439	0.04592	1.48980	196	168.0	9.4178
14	0.80558	1.7223	0.14883	0.11667	1.00000	120	168.0	15.7642
15	0.46110	3.2254	0.09671	0.04762	1.43810	210	168.0	8.9564
16	0.74869	1.9622	0.14615	0.04615	1.26923	130	168.0	17.4466
17	0.78115	2.2048	0.16038	0.10687	1.00000	131	168.0	23.1804
18	0.95861	3.5652	0.17435	0.10435	1.00000	115	168.0	22.8550
19	0.63620	5.4804	0.13676	0.03352	0.92737	179	168.0	19.8878
20	0.69958	0.7595	0.13536	0.06250	1.00000	192	168.0	11.6029
21	0.74050	1.1576	0.15381	0.06931	0.91584	202	168.0	15.4222
22	0.79637	3.9536	0.19173	0.10970	1.00000	237	168.0	20.2711
23	0.75736	1.9153	0.12846	0.15978	1.00000	363	168.0	15.2617
24	0.84879	3.2531	0.01625	0.05298	1.19205	453	168.0	18.2489

The redefined variables reveal some features of the data that were not originally apparent. For example, the RELCHECK variable shows that establishment 12 has a very high turnover rate while establishments 3 and 7 have low turnover rates. Now perform the following regression:

```
proc reg;
    model relman = relocc relcheck relcom relwings relcap hours
                   rooms / vif;
run;
```

The purpose of keeping the ROOMS variable in the model is to find out whether there are economies of scale, that is, if man-hour requirements per room decrease for larger establishments. Output 4.4 shows the results of the regression.

Output 4.4
Regression
with
Redefined
Variables

```
                               The REG Procedure
                                Model: MODEL1
                          Dependent Variable: RELMAN

                               Analysis of Variance

                                   Sum of         Mean
     Source              DF       Squares        Square    F Value    Pr > F

     Model                7     775.01435     110.71634       1.76    0.1659
     Error               16    1008.44749      63.02797
     Corrected Total     23    1783.46183


             Root MSE                 7.93902     R-Square     0.4346
             Dependent Mean          19.96950     Adj R-Sq     0.1872
             Coeff Var               39.75571


                               Parameter Estimates

                          Parameter      Standard                          Variance
     Variable      DF      Estimate         Error    t Value   Pr > |t|    Inflation

     Intercept      1      36.88142      21.07922       1.75     0.0993            0
     RELOCC         1       1.66891      10.01957       0.17     0.8698      2.75373
     RELCHECK       1       1.58059       1.08598       1.46     0.1649      2.09438
     RELCOM         1     -12.94336      25.50015      -0.51     0.6187      1.68052
     RELWINGS       1      20.20721      26.22190       0.77     0.4522      1.22610
     RELCAP         1     -16.02822      13.56252      -1.18     0.2546      1.52130
     ROOMS          1      -0.01943       0.01818      -1.07     0.3010      1.63477
     HOURS          1      -0.02174       0.03670      -0.59     0.5619      1.73989
```

You can see that the multicollinearity has been decreased but the regression is not statistically significant. In other words, there is little evidence that factors other than size affect manpower requirements. It is, however, possible that variable selection may reveal that some of these variables are useful in determining manpower requirements (see Section 4.4).

4.3.2 Multivariate Structure: Principal Component Regression

The principal component variables may themselves be used as independent variables in a regression analysis. Such an analysis is called principal component regression. Since the components are uncorrelated, there is no multicollinearity in the regression, and you can easily determine the important coefficients. Then, if the coefficients of the principal component transformations have a meaningful interpretation, the regression may shed light on the underlying regression relationships. Unfortunately, such interpretations are not always obvious.

Principal component regression is illustrated here with the BOQ data. The analysis is composed of two parts:

❑   Use PROC PRINCOMP to perform the principal component analysis.

❑   Use PROC REG to perform the regression of the dependent variable on the set of component variables.

The SAS statements for the principal component analysis are

```
proc princomp data = boq out = prin;
   var occup checkin hours common wings cap rooms;
run;
```

The option OUT=PRIN creates a data set called PRIN that contains the variables in the original data set as well as the new principal component variables, which are automatically named PRIN1 through PRIN7. The results of the PRINCOMP procedure using the independent variables from the BOQ data appear in Output 4.5.

Output 4.5
Principal
Components
Analysis

```
                                  The PRINCOMP Procedure

                              Observations          24
                              Variables              7

                              Simple Statistics

                        OCCUP           CHECKIN           HOURS           COMMON

         Mean       89.49208333       314.5129167      134.6041667       15.42375000
         StD        91.39426868       382.4113187       59.4962511       14.17183607

                              Simple Statistics

                          WINGS              CAP             ROOMS

           Mean       10.87500000      133.0416667      121.4583333
           StD        12.23790441      135.2156021      116.4231335


                              Correlation Matrix

                 OCCUP    CHECKIN    HOURS    COMMON    WINGS      CAP     ROOMS

    OCCUP       1.0000     0.8571    0.4785    0.5688    0.7668    0.9270    0.9708
    CHECKIN     0.8571     1.0000    0.4607    0.4640    0.5460    0.8452    0.8545
    HOURS       0.4785     0.4607    1.0000    0.3809    0.3736    0.4634    0.4799
    COMMON      0.5688     0.4640    0.3809    1.0000    0.6827    0.5878    0.6579
    WINGS       0.7668     0.5460    0.3736    0.6827    1.0000    0.6722    0.7581
    CAP         0.9270     0.8452    0.4634    0.5878    0.6722    1.0000    0.9785
    ROOMS       0.9708     0.8545    0.4799    0.6579    0.7581    0.9785    1.0000



                     Eigenvalues of the Correlation Matrix

                    Eigenvalue    Difference    Proportion    Cumulative

            1       5.04673909    4.32371330      0.7210        0.7210
            2       0.72302579    0.02363528      0.1033        0.8243
            3       0.69939050    0.38158104      0.0999        0.9242
            4       0.31780946    0.15910369      0.0454        0.9696
            5       0.15870578    0.10821612      0.0227        0.9922
            6       0.05048966    0.04664994      0.0072        0.9995
            7       0.00383972                    0.0005        1.0000
                              The PRINCOMP Procedure

                                  Eigenvectors

                 Prin1      Prin2      Prin3      Prin4      Prin5      Prin6      Prin7

    OCCUP      0.427129   -.016121   -.223824   -.190675   -.106859   -.773655   0.347686
    CHECKIN    0.385625   0.176465   -.402776   0.271250   0.748651   0.154482   -.002249
    HOURS      0.258063   0.832573   0.476599   -.106772   -.037258   0.015315   -.007541
    COMMON     0.321746   -.368518   0.613557   0.601401   0.056206   -.109608   0.085824
    WINGS      0.362662   -.369182   0.302495   -.707400   0.264716   0.264551   0.014167
    CAP        0.420433   0.003662   -.256780   0.123786   -.510810   0.541268   0.433666
    ROOMS      0.436405   -.057519   -.163207   0.035291   -.304228   -.048847   -.826692
```

The first portion of the output provides the descriptive statistics: the means, standard deviations, and correlation coefficients. You can easily see the very high correlation coefficients among OCCUP, CAP, and ROOMS, but other correlations are also quite large.

The second portion provides information on the eigenvalues of the correlation matrix. The eigenvalues are identified by column headings. Since there are seven variables, there are also seven eigenvalues that are arranged from high to low and identified as Prin1–Prin7. As noted, these eigenvalues are the same as those obtained by the COLLINOINT option in the MODEL statement in PROC REG (see Output 4.2).

As noted, the sum of the variances of the original seven variables, as well as the sum of variances of the seven new principal component variables (the eigenvalues), is seven. This is a measure of the total variation inherent in the entire set of data. The first principal component shows a very large variance (5.04), the second and third have modest variances (0.72 and 0.70), and the others have very small variances (less than 0.5). Remember that an eigenvalue of zero implies exact linear dependency and the two very small eigenvalues indicate two near collinearities.

The Difference column gives the differences between adjacent eigenvalues. This statistic shows the rate of decrease in variances of the principal components.

The proportion of total variation accounted for by each of the components is obtained by dividing each of the eigenvalues by the total variation. These quantities are given in the Proportion column. You can see that the first component accounts for $5.04/7 = 72\%$ of the total variation, a result that is typical when a single factor, in this case the size of the establishment, is a common factor in the variability among the original variables.

The cumulative proportions printed in the Cumulative column indicate, for each component, the proportion of the total variation of the original set of variables explained by all components up to and including that one. For example, 92% of the total variation in the seven variables is explained by only three components. This is another indication that the original set of variables contains redundant information.

The columns of the final portion of the output give the eigenvectors for each of the principal components. These coefficients, which relate the components to the original variables listed on the left, are scaled so that their sum of squares is unity. This allows for finding which of the original variables dominate a component.

The coefficients of the first principal component show a positive relationship with all variables, with somewhat larger contributions from OCCUP, CAP, and ROOMS. As expected, this component measures the size of the establishment. The second component is dominated by HOURS. This shows that among these establishments there is variation in operating hours that is relatively independent of size. The third component is dominated by COMMON and is also somewhat a function of HOURS and, negatively, of CHECKIN. This component indicates variability among the establishments that reflects large common areas and longer hours of operation but a relatively low number of check-ins. This component may differentiate between establishments that have more transients and those that have more long-term residents. Interpretation of components having small eigenvalues, such as components four through seven, is not usually useful, although such components may reveal data anomalies (see Output 4.7).

Next, you can perform a regression of MANH on the principal components. Remember that the implementation of the OUT=PRIN option in the PROC PRINCOMP step created a data set called PRIN that contains all the original variables as well as the seven principal components which are identified as variables Prin1–Prin7. Now perform the following regression:

```
proc reg data = prin;
   model manh = prin1 - prin7 / ss2;
run;
```

The results appear in Output 4.6.

Output 4.6
Principal
Components
Regression

```
                          The REG Procedure
                           Model: MODEL1
                       Dependent Variable: MANH

                          Analysis of Variance

                                 Sum of         Mean
     Source            DF       Squares       Square    F Value   Pr > F

     Model              7      87497673     12499668     154.32   <.0001
     Error             16       1295987        80999
     Corrected Total   23      88793659


                 Root MSE            284.60353    R-Square     0.9854
                 Dependent Mean     2050.00708    Adj R-Sq     0.9790
                 Coeff Var            13.88305


                          Parameter Estimates

                       Parameter     Standard
     Variable    DF     Estimate        Error    t Value   Pr > |t|   Type II SS

     Intercept    1    2050.00708     58.09445     35.29    <.0001     100860697
     Prin1        1     827.09516     26.41624     31.31    <.0001      79405328
     Prin2        1      40.58272     69.79101      0.58    0.5690         27388
     Prin3        1    -470.67262     70.96047     -6.63    <.0001       3563571
     Prin4        1    -173.96713    105.26717     -1.65    0.1179        221222
     Prin5        1     461.60751    148.96355      3.10    0.0069        777797
     Prin6        1   -1733.06676    264.10410     -6.56    <.0001       3487875
     Prin7        1     405.07458    957.69358      0.42    0.6779         14491
```

The statistics for the overall regression (SS Model and SS Error) are, by definition, the same as for the regression with the seven original variables (see Output 4.1). However, the statistics for the variables in the model tell a different story. Since the components are uncorrelated, the VIFs (not printed) are all unity. You can also verify that the Type II sums of squares add to the model sum of squares. According to the t statistics for the parameter estimates, there appear to be four components of importance for estimating man-hour requirements.[3] The most important component is Prin1, followed in importance by Prin3 and Prin6, and to a lesser degree Prin5.

Prin1 is clearly the component associated with size of establishment and accounts for over 90% of the variation explained by the model (Type II SS/Model SS). The second component apparently does nothing; that is, the hours of operation apparently have no effect on manpower requirements. Component three was associated with common areas, with hours, and (negatively) with check-ins. The negative coefficient indicates lower man-hour requirements for establishments having larger common areas and longer desk hours and lower turnover. In other words, establishments serving primarily longer-term residents require less manpower.

---

[3] You should not use the p values literally, but the magnitudes of the p values can be used to indicate relative importance of the coefficients.

Because principal components with very small variances are usually not useful, strong relationships of such components with the dependent variable are an apparent contradiction and may very well indicate data anomalies. You can investigate this by plotting the component variables against the dependent variable. In this example, components three and six have small variances and appear important and are therefore logical candidates for such plots. If you are using PROC REG with the LINEPRINTER option, enter the following statements:[4]

```
plot prin3*manh prin6*manh  / hplots = 2 vplots = 2;
run;
```

Otherwise, use PROC PLOT as follows:

```
proc plot  data = prin vpercent = 50 hpercent = 50 ;
   plot prin3*manh  prin6*manh ;
run;
```

The HPLOT and VPLOT as well as the HPERCENT and VPERCENT options produce side-by-side plots one-half-page long. The results, using PROC REG interactively, appear in Output 4.7.

Output 4.7
Plots of
Components
Three and
Six

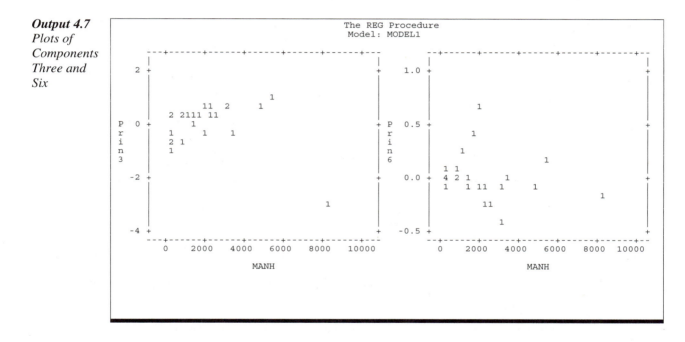

You can see that the relationships of both Prin3 and Prin6 with MANH are dominated by one observation, which can be identified as number 24. Remember that observation 24 was unusual in having very high turnover and small common areas, which have relatively large coefficients for component three. The relationship of Prin6 to MANH is also affected by observation 24, but the reason for this is not clear.

Principal components that have large eigenvalues can also be used to re-create values of the original variables reflecting only the variation explained by these components. This is a biased estimation procedure and is illustrated in Section 4.5.1.

---

[4] If you are not in the line printer mode you cannot use the HPLOTS and VPLOTS options, and you will get two separate high-resolution plots.

In summary, principal component regression has helped somewhat to interpret the structure of the regression, although the interpretation is not clear, which is common with this type of analysis. Slightly more interpretable results may be obtained by using rotated principal components, which are available in the FACTOR procedure. Presentation of such methods is beyond the scope of this book.

4.4 Variable Selection

When a number of variables in a regression analysis do not appear to contribute significantly to the predictive power of the model, it is natural to try to find a suitable subset of important or useful variables. Regression methodology assumes that you have specified the appropriate model, but in many cases theory or intuitive reasoning does not supply such a model. In such situations, it is customary to start with a model containing a number of candidate variables and then use an automated procedure that uses information from the regression analysis to select a suitable subset of variables for a final model.

An optimum subset model is one that, for a given number of variables, produces the minimum error sum of squares, or, equivalently, the maximum R^2. Essentially, the only way to ensure finding optimum subsets is to examine all possible subsets. This procedure, for m independent variables, requires computing statistics for $m!/[p!(m-p)!]$ regression equations for subsets of size p from an m-variable model, or 2^m equations for finding optimum subsets for all subset sizes. Fortunately, optimal search procedures, highly efficient algorithms, and high-speed computing capabilities make such a procedure feasible for models with a moderate number of variables. Such a procedure is implemented by an option in PROC REG, and it is normally recommended for models containing fewer than 25 variables. This is a somewhat arbitrary limit and depends on the speed of the computer used. However, remember that, for example, a 26-variable model will essentially require 26 times more computing than a 25-variable one. This method is presented in Section 4.4.1, "The R-Square Selection Method."

Popular alternatives to the guaranteed optimum subset selection are the step-type procedures that add or delete variables one at a time until, by some criterion, a reasonable stopping point is reached. These procedures do not guarantee finding optimum subsets, but they work quite well in many cases and are especially useful for models with many variables. A number of step-type procedures may be implemented as options in PROC REG and are presented in Section 4.4.3, "Other Variable Selection Procedures."

Because these selection methods often produce a bewildering array of results, statistics have been developed to assist in choosing suitable subset models. The use of such statistics is presented in Section 4.4.2, "Choosing Useful Models."

Before continuing, it is necessary to point out that although advances in computing power have made variable selection easy to do, it is not always appropriate. In fact, variable selection may produce misleading results and may lead to a selected set that is not optimal for the population (Freund and Wilson 1998, Section 6.5). Additional comments on this issue are presented in Section 4.6, "Summary."

4.4.1 The R-Square Selection Method

Guaranteed optimum subsets are obtained by using the MODEL statement option
SELECTION=RSQUARE in PROC REG. You can use this method for the BOQ data as follows:

```
proc reg dat = boq;
   model manh = occup checkin hours common wings cap rooms /
       selection = rsquare  best = 4 cp;
run;
```

Two additional options are specified here as follows:

BEST=4   specifies that only the best four (smallest error mean square) models for each subset
            size are to be printed. This option prevents excessive output.

CP   specifies the printing of the Mallows C(P) statistic (denoted by C(P) in the output)
            for each subset. This is the most popular of several statistics used to aid selection of
            a final model (see Section 4.4.2).

The results appear in Output 4.8.

Output 4.8
Regression
for BOQ
Data Using
R-SQUARE
Selection

```
                            The REG Procedure
                            Model: MODEL1
                         R-Square Selection Method
                 Regression Models for Dependent Variable: MANH

   Number in
    Model      R-Square      C(p)       Variables in Model

       1        0.9619     21.7606      OCCUP
       1        0.8888    101.8743      ROOMS
       1        0.8149    182.8633      CHECKIN
       1        0.7920    207.9875      CAP
   --------------------------------------------------------------------------
       2        0.9765      7.8089      OCCUP CHECKIN
       2        0.9645     20.8704      OCCUP CAP
       2        0.9634     22.1085      OCCUP ROOMS
       2        0.9629     22.7211      OCCUP WINGS
   --------------------------------------------------------------------------
       3        0.9839      1.7003      OCCUP CHECKIN CAP
       3        0.9803      5.6299      OCCUP CHECKIN ROOMS
       3        0.9767      9.5345      OCCUP CHECKIN COMMON
       3        0.9766      9.6912      OCCUP CHECKIN WINGS
   --------------------------------------------------------------------------
       4        0.9851      2.3204      OCCUP CHECKIN COMMON CAP
       4        0.9845      2.9396      OCCUP CHECKIN CAP ROOMS
       4        0.9839      3.6414      OCCUP CHECKIN COMMON ROOMS
       4        0.9839      3.6868      OCCUP CHECKIN WINGS CAP
   --------------------------------------------------------------------------
       5        0.9854      4.0091      OCCUP CHECKIN COMMON WINGS CAP
       5        0.9851      4.3159      OCCUP CHECKIN HOURS COMMON CAP
       5        0.9851      4.3195      OCCUP CHECKIN COMMON CAP ROOMS
       5        0.9846      4.8286      OCCUP CHECKIN WINGS CAP ROOMS
   --------------------------------------------------------------------------
       6        0.9854      6.0024      OCCUP CHECKIN HOURS COMMON WINGS CAP
       6        0.9854      6.0059      OCCUP CHECKIN COMMON WINGS CAP ROOMS
       6        0.9851      6.3146      OCCUP CHECKIN HOURS COMMON CAP ROOMS
       6        0.9847      6.8219      OCCUP CHECKIN HOURS WINGS CAP ROOMS
   --------------------------------------------------------------------------
       7        0.9854      8.0000      OCCUP CHECKIN HOURS COMMON WINGS CAP ROOMS
```

The output from the SELECTION=RSQUARE option provides, for each subset size, the variables
included in the four best models, listed in order of decreasing R-Square, along with the R-Square
and C(P) statistics. You can see that R-Square remains virtually unchanged down to subsets of
size three, with all four selections having nearly equal R-Square values. Additional considerations
in this selection are presented in Section 4.4.2.

Some additional options that are useful for problems with many variables are as follows:

INCLUDE=*n*   specifies that the first *n* independent variables in the MODEL statement are to be included in all subset models.

START=*n*   specifies that only subsets of *n* or more variables are to be considered. This number includes the variables specified by the INCLUDE option.

STOP=*n*   specifies that subsets of no more than *n* variables are to be considered. This number also includes the variables specified by the INCLUDE option.

B   specifies that the regression coefficients are printed for each subset model. This option should be used sparingly as it will produce a large amount of output. A more efficient way of getting coefficient estimates is to use information from Output 4.8 to choose interesting models and obtain the coefficient estimates by repeated MODEL statements or, interactively, with ADD and DELETE statements.

4.4.2 Choosing Useful Models

An examination of the R-Square values in Output 4.8 does not reveal any obvious choices for selecting a most useful subset model. A number of other statistics have been developed to aid in making these choices and 12 of these are available as additional options with the RSQUARE option. Among these, the most frequently used is the C(P) statistic, proposed by Mallows (1973). This statistic is a measure of total squared error for a subset model containing *p* independent variables. The total squared error is a measure of the error variance plus the bias introduced by not including variables in a model. It may, therefore, indicate when variable selection is deleting too many variables. The C(P) statistic is computed as follows:

$$C(P) = (SSE(p)/MSE) - (N - 2p) + 1$$

where

MSE   is the error mean square for the full model (or some other estimate of pure error)

SSE(*p*)   is the error sum of squares for the subset model containing *p* independent variables (not including the intercept)[5]

N   is the total sample size.

For any given number of selected variables, larger C(P) values indicate equations with larger error mean squares. For any subset model for which C(P)>(*p*+1), there is evidence of bias due to the deletion of important variables for a model. On the other hand, if there are values of C(P)<(*p*+1), the full model is said to be overspecified; that is, it contains too many variables.

Mallows recommends that C(P) be plotted against *p*, and further recommends selecting that subset size where the minimum C(P) first approaches (*p*+1), starting from the full model. The magnitudes of differences in the C(P) statistic between the optimum and near optimum models for each subset size are also of interest.

---

[5] In the original presentation of the C(P) statistic (Mallows 1973), the intercept coefficient is also considered as a candidate for selection, so that in that presentation the number of variables in the model is one more than what is defined here and results in the +1 elements in the equations. As implied in the discussion of the COLLIN option, allowing the deletion of the intercept is not normally useful.

You can obtain a high-resolution C(P) plot directly after having run PROC REG with the RSQUARE option by using the following :

```
plot cp.*in. np.*in. / overlay vaxis = 0 to 15 by 5;
    symbol1 v='C' c=black;
    symbol2 v='star' c=black;
run;
```

The specifications for the plot are

CP.* IN.   provides the plot of C(P) against the number of variables.

NP.*IN.   provides the reference line C(P) = number of variables plus one.

VAXIS option   deletes from the plot the very large C(P) values, which are of no interest.

SYMBOL   specifies that the C(P) values are plotted with the "C" symbol and the
statement   reference line values are plotted with "*".

The plot is shown in Output 4.9.

Output 4.9
Interactive
C(p) Plot

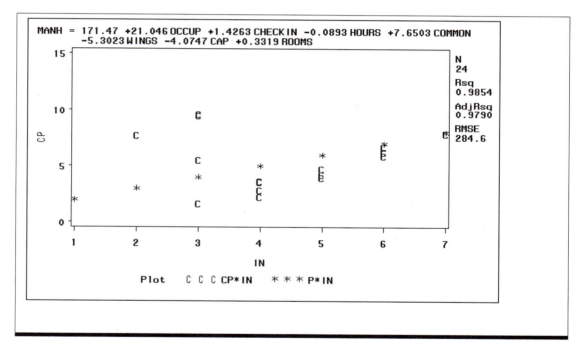

Alternately you can obtain such a plot by using the OUTEST option with SELECTION= RSQUARE and use PROC GPLOT. You invoke PROC REG as follows:

```
proc reg outest = est;
    model manh = occup checkin hours common wings cap rooms /
        selection = rsquare CP best = 4;
    proc print data=est;
run;
```

You can examine the resulting data set using PROC PRINT. Output 4.10 shows the results.

Output 4.10
Output Data Set from R-Square Selection

Obs	_MODEL_	_TYPE_	_DEPVAR_	_RMSE_	Intercept	OCCUP	CHECKIN	HOURS	COMMON
1	MODEL1	PARMS	MANH	660.841	119.601	.	.	.	.
2	MODEL1	PARMS	MANH	848.463	589.743	.	4.59794	.	.
3	MODEL1	PARMS	MANH	896.178	330.003	.	.	.	.
4	MODEL1	PARMS	MANH	565.654	168.046	.	1.81601	.	.
5	MODEL1	PARMS	MANH	608.909	70.808	.	.	.	.
6	MODEL1	PARMS	MANH	649.474	294.833	.	3.64455	.	.
7	MODEL1	PARMS	MANH	477.343	117.588	.	1.92858	.	.
8	MODEL1	PARMS	MANH	548.021	147.328	.	2.10226	.	.
9	MODEL1	PARMS	MANH	563.885	199.307	.	.	.	-25.2908
10	MODEL1	PARMS	MANH	455.909	198.828	.	1.71925	.	-16.9887
11	MODEL1	PARMS	MANH	471.375	107.419	-0.99790	2.00464	.	.
12	MODEL1	PARMS	MANH	488.188	116.388	.	1.97497	.	.
13	MODEL1	PARMS	MANH	434.089	201.363	-1.32493	1.77891	.	-20.3421
14	MODEL1	PARMS	MANH	462.829	202.148	.	1.80661	.	-18.2391
15	MODEL1	PARMS	MANH	466.532	137.923	.	1.69643	0.59676	-17.3811
16	MODEL1	PARMS	MANH	444.049	203.275	-1.27527	1.82950	.	-20.9723
17	MODEL1	PARMS	MANH	444.228	131.948	-1.33282	1.75326	0.68028	-20.8094
18	MODEL1	PARMS	MANH	474.301	142.597	.	1.78383	0.58331	-18.6157
19	MODEL1	PARMS	MANH	455.167	134.968	-1.28377	1.80351	0.66915	-21.4226

Obs	WINGS	CAP	ROOMS	MANH	_IN_	_P_	_EDF_	_RSQ_	_CP_
1	.	.	15.7569	-1	1	2	23	0.88951	27.4820
2	.	.	.	-1	1	2	23	0.81787	58.9192
3	.	12.9504	.	-1	1	2	23	0.79681	68.1609
4	.	.	10.6204	-1	2	3	22	0.92257	14.9768
5	.	-10.0640	27.0936	-1	2	3	22	0.91027	20.3718
6	54.8568	.	.	-1	2	3	22	0.89792	25.7924
7	.	-11.0267	22.7231	-1	3	4	21	0.94737	6.0961
8	23.3843	.	7.9761	-1	3	4	21	0.93062	13.4419
9	.	-13.3330	32.7806	-1	3	4	21	0.92655	15.2299
10	.	-13.1181	27.0176	-1	4	5	20	0.95427	5.0652
11	.	-12.2805	24.9040	-1	4	5	20	0.95112	6.4497
12	4.2592	-10.4644	21.6242	-1	4	5	20	0.94757	8.0071
13	.	-15.1956	30.7609	-1	5	6	19	0.96062	4.2810
14	9.4346	-12.0264	24.8996	-1	5	6	19	0.95523	6.6450
15	.	-13.1176	26.9655	-1	5	6	19	0.95451	6.9606
16	5.7038	-14.4577	29.3402	-1	6	7	18	0.96096	6.1313
17	.	-15.2074	30.7238	-1	6	7	18	0.96093	6.1452
18	9.3826	-12.0319	24.8604	-1	6	7	18	0.95546	8.5452
19	5.6192	-14.4803	29.3248	-1	7	8	17	0.96126	8.0000

As you can see, this data set contains a number of statistics. The estimated coefficients are identified by the names of the independent variables. For the C(p) plot, plot the variable _CP_ against _IN_ for the C(P) values and _P_ against _IN_ for the reference line.

```
proc gplot;
    symbol1 v = 'C' c=black;
    symbol2 v = star c=black l=1 I=join;
    plot _cp_*_in_=1 _p_*_in_=2 / overlay vaxis= 0 to 15 by 5;
run;
```

The plot, which is not reproduced here, will look like the one in Output 4.9 except for the line joining the reference points.

The pattern of C(P) values for this example is quite typical for situations where multicollinearity is serious. Starting with _IN_=7, they initially become smaller than (p+1), as fewer variables are included, but eventually start to increase. In other words, the residual mean square initially decreases as variables are deleted. In this plot, there is a definite "corner" at _IN_=3, where the C(P) values increase rapidly with smaller subset sizes. Hence, a model with three variables appears to be a good choice.

A second criterion is the difference in C(P) between the optimum and second optimum subset for each subset size. In the above models with four or more variables, these differences are very small, which implies that the multicollinearity allows interchange of variables without affecting the fit of the model. However, that difference is larger with three variables, implying that the degree of multicollinearity has decreased. You can now estimate the best three-variable model using the information from Output 4.8 as follows:

```
proc reg data = boq;
    model manh = occup checkin cap / vif;
run;
```

Results of the regression appear in Output 4.11.

Output 4.11
Regression
for BOQ
Data, Best
Three
Variable
Model

```
                            The REG Procedure
                               Model: MODEL1
                         Dependent Variable: MANH

                            Analysis of Variance

                                  Sum of        Mean
   Source              DF        Squares      Square    F Value    Pr > F

   Model                3       87359949    29119983     406.22    <.0001
   Error               20        1433710       71686
   Corrected Total     23       88793659


               Root MSE              267.74149    R-Square     0.9839
               Dependent Mean       2050.00708    Adj R-Sq     0.9814
               Coeff Var              13.06052


                            Parameter Estimates

                        Parameter      Standard                          Variance
   Variable     DF       Estimate         Error    t Value   Pr > |t|   Inflation

   Intercept     1      207.86486      78.28539       2.66     0.0152           0
   OCCUP         1       20.67163       1.75123      11.80     <.0001     8.21908
   CHECKIN       1        1.43624       0.29366       4.89     <.0001     4.04615
   CAP           1       -3.45397       1.14110      -3.03     0.0067     7.63825
```

Note: Since this model has been specified by the data, the *p* values cannot be used literally but are useful for determining the relative importance of the variables.

It does appear that more turnovers (CHECKIN) require additional manpower. The negative *partial* coefficient for CAP may reflect lower man-hours for a larger proportion of vacant rooms.

A number of other statistics are available to assist in choosing subset models. Some are relatively obvious, such as the residual mean square or standard deviations, while others are related to R-Square, with some providing adjustments for degrees of freedom or scaling preferences. They are all essentially equivalent, although some have different theoretical justification. A number of these are available as options with the selection methods described in this section. Keywords and literature references for these options are provided in the PROC REG chapter in the *SAS/STAT User's Guide*.

For those who find the results of the SELECTION=RSQUARE method somewhat bewildering, PROC REG provides for two less-bewildering but also less-informative methods for selecting optimum subsets. These selection methods are SELECTION=ADJRSQ and SELECTION=CP. For each of these, PROC REG prints a total of *n* sets of subset models judged optimum according to the respective criterion (ADJ R-SQ or C(P)), where *n* is specified by the BEST=*n* option.

Now, for a given subset size, there is a monotonic relationship among all three selection criteria, but this is not true when they are compared across different subset sizes. Thus, for example, the SELECTION=CP BEST=4 option produces the four models with the smallest C(P) values *regardless of subset size*. From Output 4.8, you can verify that this is the optimum three-variable model followed by the three best four-variable models. Thus, this selection method provides a compact output for a few optimum models according to this criterion, but you will not, for example, have the comprehensive summary of selection results provided by the plot in Output 4.10.

In the same manner, SELECTION=ADJRSQ provides *n* models with the largest adjusted R-Square values (see Section 2.3, "A Model with Several Independent Variables").

4.4.3 Other Variable Selection Procedures

Several step-type selection procedures are available as alternatives to the R-Square selection method. Starting with some given model, a step can do any of the following:

❑   add a variable to the model

❑   delete a variable from the model

❑   exchange a variable in the model for one that is not in the model.

Steps continue until a specified stopping point has been reached.

Five different step-type selection methods are available as MODEL options in the PROC REG step. Each is implemented with SELECTION=KEYWORD, where the keyword is given in parentheses in the descriptions that follow. Additional options for these methods are detailed after the descriptions. Only one selection option is available with any MODEL statement, but several MODEL statements may be used each with a different selection option. The selection methods are as follows:

Forward selection
(FORWARD or F)

begins by finding the variable that produces the optimum one-variable model. In the second step, the procedure finds the variable that, when added to the already chosen variable, results in the largest reduction in the residual sum of squares (largest increase in R^2). The third step finds the variable that, when added to the two already chosen, gives the minimum residual sum of squares (maximum R^2). The process continues until no variable considered for addition to the model provides a reduction in sum of squares considered statistically significant at a level specified by the user (see SLE specification later in this section). An important feature of this method is that once a variable has been selected, it stays in the model.

Backward elimination
(BACKWARD or B)

begins by computing the regression with all independent variables specified in the MODEL statement. The procedure deletes from that model the variable whose coefficient has the largest p value (smallest partial F value). The resulting equation is examined for the variable now contributing the least, which is then deleted, and so on. The procedure stops when all coefficients remaining in the model are statistically significant at a level specified by the user (see SLS specification later in this section). With this method, once a variable has been deleted, it is deleted permanently.

Stepwise selection
(STEPWISE)

begins like forward selection, but after a variable has been added to the model, the resulting equation is examined to see if any coefficient has a sufficiently large p value (see SLS specification later in this section) to suggest that a variable should be dropped. This procedure continues until no additions or deletions are indicated according to significance levels (SLE and SLS) chosen by the user.

Maximum R-Square improvement
(MAXR)

begins by selecting one- and two-variable models as in forward selection. At that point, the procedure examines all possible pairwise interchanges with the variables not in the model. Among all interchanges that increase R^2, that interchange resulting in the largest increase in R^2 is implemented. This process is repeated until no pairwise interchange improves the model. At this point, a third variable is selected as in forward selection. Then the interchanging process is repeated, and so on. This method usually requires more computer time than the other three, but it also tends to have a better chance of finding more nearly optimum models. In addition, the maximum R-Square improvement method often produces a larger number of equations, a feature that may be of some value in the final evaluation process.

Minimum R-Square improvement
(MINR)

is similar to maximum R-Square improvement, except that interchanges are implemented for those with minimum improvement. Since interchanges are not implemented when R^2 is decreased, the final results are quite similar to those of the maximum R-Square improvement method, except that a larger number of equations may be examined.

A number of options are available to provide greater control over the selection procedures. One option available for all procedures is INCLUDE=n, which specifies that the first n independent variables in the MODEL statement are to be kept in the model at all times.

For the FORWARD, BACKWARD, and STEPWISE methods, you may specify desired significance levels for stopping the addition or elimination of variables as follows:

SLE = .*xxx*   specifies the significance level for stopping the addition of variables in the forward selection mode. If not specified, the default is 0.50 for FORWARD and 0.15 for STEPWISE.

SLS = .*xxx*   specifies the significance level for stopping the backward elimination mode. If not specified, the default is 0.10 for BACKWARD and 0.15 for STEPWISE.

The smallest permissible value for SLS is 0.0001, which almost always ensures that the final equation obtained by backward elimination contains only one variable, while the maximum SLE of 0.99 usually includes all variables when forward selection has stopped. It is again important to note that, since variable selection is an exploratory rather than confirmatory analysis, the SLE and SLS values do not have the usual interpretation as probabilities of erroneously rejecting the null hypothesis of the nonexistence of the coefficients in any selected model.

MAXR and MINR selection do not use significance levels, but you may specify the starting and stopping of these procedures with these options:

START=*s*   specifies that the interchanging procedure starts with the first *s* variables in the MODEL statement.

STOP=*s*   specifies that the procedure stops when the best *s*-variable model has been found.

For MAXR and MINR, the INCLUDE option works as it does with RSQUARE. That is, if the INCLUDE option is used in addition to the START or STOP option, the INCLUDE option overrides the START or STOP specification. For example, if you use START=3 and INCLUDE=2, the selection starts with the first three variables, but the first two may not be candidates for deletion.

Because the outputs from the step-type selection procedures are quite lengthy, they are illustrated here only with the SELECTION=FORWARD option using the redefined variables of the BOQ data. Use these SAS statements:

```
proc reg data = boq;
   model relman = relocc relcheck relcom relwings relcap
      hours rooms / selection = f ;
```

The results appear in Output 4.12.

Output 4.12
Forward
Selection for
Redefined
Variables

```
                        BOQ DATA, MYERS P 145
                            OMIT OBS 23

                          The REG Procedure
                            Model: MODEL1
                      Dependent Variable: RELMAN

                       Forward Selection: Step 1

        Variable RELCAP Entered: R-Square = 0.1849 and C(p) = 3.0633


                          Analysis of Variance

                                   Sum of          Mean
        Source             DF      Squares        Square    F Value   Pr > F

        Model              1      329.82588     329.82588     4.99    0.0360
        Error             22     1453.63595      66.07436
        Corrected Total   23     1783.46183


                         Parameter     Standard
            Variable      Estimate       Error    Type II SS  F Value  Pr > F

            Intercept     46.25587     11.88177   1001.39207   15.16   0.0008
            RELCAP       -25.15406     11.25854    329.82588    4.99   0.0360

                     Bounds on condition number: 1, 1
--------------------------------------------------------------------------------


                                                                 Continued
```

```
                         Forward Selection: Step 2

          Variable RELCHECK Entered: R-Square = 0.2908 and C(p) = 2.0666

                           Analysis of Variance

                                    Sum of          Mean
Source                   DF         Squares         Square    F Value   Pr > F

Model                     2       518.70510       259.35255      4.31    0.0271
Error                    21      1264.75674        60.22651
Corrected Total          23      1783.46183


                         BOQ DATA, MYERS P 145
                            OMIT OBS 23

                          The REG Procedure
                            Model: MODEL1
                       Dependent Variable: RELMAN

                        Forward Selection: Step 2

                   Parameter     Standard
    Variable       Estimate       Error     Type II SS  F Value  Pr > F

    Intercept      44.18015      11.40420    903.88394   15.01   0.0009
    RELCHECK        1.30004       0.73411    188.87921    3.14   0.0911
    RELCAP        -25.90676      10.75719    349.31401    5.80   0.0253

          Bounds on condition number: 1.0016, 4.0063
---------------------------------------------------------------------------------
                         Forward Selection: Step 3

          Variable ROOMS Entered: R-Square = 0.3868 and C(p) = 1.3521
                           Analysis of Variance

                                    Sum of          Mean
Source                   DF         Squares         Square    F Value   Pr > F

Model                     3       689.79536       229.93179      4.20    0.0185
Error                    20      1093.66647        54.68332
Corrected Total          23      1783.46183


                   Parameter     Standard
    Variable       Estimate       Error     Type II SS  F Value  Pr > F

    Intercept      39.37033      11.20177    675.49121   12.35   0.0022
    RELCHECK        1.53970       0.71251    255.35652    4.67   0.0430
    RELCAP        -18.83125      11.00307    160.17146    2.93   0.1025
    ROOMS          -0.02562       0.01448    171.09027    3.13   0.0922

          Bounds on condition number: 1.1961, 10.168
---------------------------------------------------------------------------------
                         Forward Selection: Step 4

          Variable RELWINGS Entered: R-Square = 0.4115 and C(p) = 2.6512

                           Analysis of Variance

                                    Sum of          Mean
Source                   DF         Squares         Square    F Value   Pr > F

Model                     4       733.97347       183.49337      3.32    0.0319
Error                    19      1049.48837        55.23623
Corrected Total          23      1783.46183


                   Parameter     Standard
    Variable       Estimate       Error     Type II SS  F Value  Pr > F

    Intercept      33.97124      12.77478    390.60679    7.07   0.0155
    RELCHECK        1.55392       0.71628    259.96679    4.71   0.0429
    RELWINGS       21.68388      24.24630     44.17811    0.80   0.3823
    RELCAP        -16.36282      11.39781    113.84070    2.06   0.1674
    ROOMS          -0.02248       0.01497    124.50074    2.25   0.1497

          Bounds on condition number: 1.2656, 18.91
---------------------------------------------------------------------------------
                                                       Continued
```

*Output 4.12
(Continued)
Forward
Selection for
Redefined
Variables*

```
No other variable met the 0.5000 significance level for entry into the model.

                         Summary of Forward Selection

            Variable      Number    Partial     Model
   Step     Entered       Vars In   R-Square    R-Square    C(p)      F Value    Pr > F

    1       RELCAP           1       0.1849      0.1849     3.0633     4.99      0.0360
    2       RELCHECK         2       0.1059      0.2908     2.0666     3.14      0.0911
    3       ROOMS            3       0.0959      0.3868     1.3521     3.13      0.0922
    4       RELWINGS         4       0.0248      0.4115     2.6512     0.80      0.3823
```

For each step, the output describes the action taken. In this case, Step 1 selects RELCAP, and an abbreviated output summarizes the resulting model. In the same fashion, Step 2 adds RELCHECK, Step 3 adds ROOMS, and Step 4 adds RELWINGS. The note at the end of Step 4 indicates that no other variables are added for the stated SLE. The Bounds on condition number is provided to detect possible round-off error. Values in the millions may indicate the presence of a round-off error (Berk 1977), which we obviously do not have here. At the end of the selection is a summary of the selection process.

Remember that the *p* values cannot be taken literally, but you can use them as a measure of relative importance. In the three-variable model, RELCHECK appears to have some importance and the other variables except RELWINGS have *p* values sufficiently small to indicate that they should not be ignored. Also, the signs of the coefficients do make sense, in that manpower requirements increase with greater turnover and wings but decrease with overall size (economy of scale) and fewer unused rooms.

This example provides an illustration of the fact that the step-type procedures may or may not produce optimum selections.[6] You can verify this fact by implementing the SELECTION=RSQUARE and CP options, which produces the results in Output 4.13.

---

[6] In the model using the original variables, all step-type procedures produce the optimum subsets.

Output 4.13
R-Square
Selection
with
Redefined
Variables

```
                        BOQ DATA, MYERS P 145
                             OMIT OBS 23

                          The REG Procedure
                           Model: MODEL1
                      Dependent Variable: RELMAN

                        R-Square Selection Method

    Number in
     Model    R-Square      C(p)     Variables in Model

       1       0.1849     3.0633     RELCAP
       1       0.1558     3.8866     RELOCC
       1       0.1460     4.1640     ROOMS
       1       0.1123     5.1192     RELWINGS
    ---------------------------------------------------------------------------
       2       0.2970     1.8934     RELCHECK ROOMS
       2       0.2908     2.0666     RELCHECK RELCAP
       2       0.2833     2.2801     RELOCC ROOMS
       2       0.2659     2.7728     RELOCC RELCAP
    ---------------------------------------------------------------------------
       3       0.3868     1.3521     RELCHECK RELCAP ROOMS
       3       0.3477     2.4573     RELCHECK RELWINGS ROOMS
       3       0.3417     2.6265     RELCHECK RELWINGS RELCAP
       3       0.3306     2.9415     RELOCC RELCAP ROOMS
    ---------------------------------------------------------------------------
       4       0.4115     2.6512     RELCHECK RELWINGS RELCAP ROOMS
       4       0.4002     2.9718     RELCHECK RELCOM RELCAP ROOMS
       4       0.3915     3.2187     RELOCC RELCHECK RELCAP ROOMS
       4       0.3915     3.2196     RELCHECK RELCAP HOURS ROOMS
    ---------------------------------------------------------------------------
       5       0.4217     4.3642     RELCHECK RELCOM RELWINGS RELCAP ROOMS
       5       0.4172     4.4916     RELOCC RELCHECK RELWINGS RELCAP ROOMS
       5       0.4162     4.5199     RELCHECK RELWINGS RELCAP HOURS ROOMS
       5       0.4133     4.6001     RELCHECK RELCOM RELCAP HOURS ROOMS
    ---------------------------------------------------------------------------
       6       0.4336     6.0277     RELCHECK RELCOM RELWINGS RELCAP HOURS ROOMS
       6       0.4255     6.2576     RELOCC RELCHECK RELWINGS RELCAP HOURS ROOMS
       6       0.4222     6.3509     RELOCC RELCHECK RELCOM RELWINGS RELCAP ROOMS
       6       0.4136     6.5939     RELOCC RELCHECK RELCOM RELCAP HOURS ROOMS
    ---------------------------------------------------------------------------
       7       0.4346     8.0000     RELOCC RELCHECK RELCOM RELWINGS RELCAP HOURS ROOMS
```

This output shows that the optimum two-variable model includes RELCHECK and ROOMS with a C(P) value of 1.8934 while the FORWARD method picked RELCHECK and RELCAP with a C(P) value of 2.0666. Admittedly, the difference is not large, but the FORWARD choice is not optimal. Actually, the BACKWARD method (not shown) does pick the optimum two-variable model but also picks the definitely nonoptimal one-variable model with ROOMS.

4.4.4 A Strategy for Models with Many Variables

Because SELECTION=RSQUARE allows you to investigate all models, it not only provides the optimum models for all subset sizes, but it also provides information on other subsets that, although not strictly optimal, may be very useful. However, the R-Square selection method simply cannot be used for problems with a large number of variables. In such cases, step-type procedures provide an alternative. An especially attractive alternative is to use both the FORWARD and BACKWARD methods with the SLE and SLS parameters set to provide the entire range (usually SLE=.99 and SLS=.0001). The closeness of agreement of the two methods provides some clues to true optimality: if they are identical it is quite likely that optimality has been achieved.

In addition, these methods may indicate approximately how many variables are needed. Then, for example, if only a few are needed, you may be able to implement the R-Square selection method for that limited number of variables.

4.5 Biased Estimation

Least-squares estimators provide unbiased estimates of parameters. That is, on the average, the estimate targets the true value of the parameter. For some situations, it may be possible to provide a biased estimator that has a smaller variance than does the unbiased estimator. The precision of a biased estimate, called the *mean squared error*, is the square of the bias plus the variance. In some cases, the mean squared error of a biased estimate may be smaller than the variance of the unbiased estimate.

Models based on biased estimation methods may be more useful than models based on variable selection for predicting the response for one sample using models estimated from another sample. One reason for this is that the results of variable selection are often unstable; hence, using the results of a variable selection based on one sample may not provide a good model for another sample from the same population. Biased estimation may also be useful for subsequent variable selection since such a procedure may delete variables because of weak relationships to the response rather than multicollinearity.

PROC REG provides options for two biased estimation methods: incomplete principal component regression and ridge regression.

4.5.1 Incomplete Principal Component Regression

As noted in Section 4.3.2, principal components provide a linear transformation of a set of variables to a new set of uncorrelated variables. Variables in the transformed set having very small variances may be considered as indicators of multicollinearity. The idea behind incomplete principal component (IPC) regression is to delete from the principal component regression one or more of the transformed variables having small variances and then convert the resulting regression to the original variables.

Using the MODEL option in PROC REG:

   PCOMIT = *list*

performs an incomplete principal component regression deleting, for each value *m* in the list, *m* principal components. This option provides no printed output; all results are provided in the OUTEST data set, which must be specified as a PROC option. This data set can be used to examine coefficients, to make various plots, and finally to compute estimated responses using PROC SCORE.

Start with the following statements:

```
proc reg data = boq outest = incpc noprint;
   model manh = occup checkin hours common wings cap rooms /
       pcomit =1 - 6  outseb ;
proc print data=  incpc;
run; quit;
```

The OUTEST data set INCPC will contain the IPC regression results; the NOPRINT option is used to omit the usual regression output, which is not needed. As given here, the PCOMIT= option requests that IPC regressions are computed that delete all possible numbers of components, that is, from one to six. Alternately you may specify PCOMIT = 1 2 3 4 5 6. The OUTSEB option requests that standard errors of the coefficients be provided in the data set. The data set INCPC is shown in Output 4.14.

Output 4.14
*Output Data
Set for IPC
Analysis*

Obs	_MODEL_	_TYPE_	_DEPVAR_	_RIDGE_	_PCOMIT_	_RMSE_	Intercept	OCCUP
1	MODEL1	PARMS	MANH	.	.	284.604	171.473	21.0456
2	MODEL1	SEB	MANH	.	.	284.604	148.862	4.2890
3	MODEL1	IPC	MANH	.	1	277.645	168.145	19.5046
4	MODEL1	IPCSEB	MANH	.	1	277.645	145.019	2.2079
5	MODEL1	IPC	MANH	.	2	516.309	165.455	4.8342
6	MODEL1	IPCSEB	MANH	.	2	516.309	269.677	0.6392
7	MODEL1	IPC	MANH	.	3	541.739	120.782	5.3739
8	MODEL1	IPCSEB	MANH	.	3	541.739	281.625	0.5831
9	MODEL1	IPC	MANH	.	4	538.394	124.376	5.0109
10	MODEL1	IPCSEB	MANH	.	4	538.394	279.856	0.4039
11	MODEL1	IPC	MANH	.	5	667.652	-365.819	3.8583
12	MODEL1	IPCSEB	MANH	.	5	667.652	300.636	0.2911
13	MODEL1	IPC	MANH	.	6	653.255	-316.007	3.8654
14	MODEL1	IPCSEB	MANH	.	6	653.255	218.783	0.2834

Obs	CHECKIN	HOURS	COMMON	WINGS	CAP	ROOMS	MANH
1	1.42632	-0.08927	7.6503	-5.3023	-4.07475	0.33191	-1
2	0.33071	1.16353	8.4384	9.4528	3.30195	6.81399	-1
3	1.42870	-0.03793	5.1972	-5.7712	-5.37391	3.20825	-1
4	0.32258	1.12889	5.9794	9.1580	1.18222	0.42040	-1
5	2.12881	0.40818	-8.2067	31.6931	1.56356	2.48111	-1
6	0.56778	2.09566	10.4837	13.5187	1.07460	0.75550	-1
7	1.22511	0.69725	-10.0375	21.7081	3.30739	3.68734	-1
8	0.21626	2.19170	10.9425	12.7899	0.35200	0.28174	-1
9	1.34851	0.38505	-2.6550	11.6521	3.46665	3.74008	-1
10	0.16199	2.14865	6.8447	5.3913	0.29857	0.27341	-1
11	0.85277	4.15540	17.7224	23.2861	2.57283	3.08027	-1
12	0.09805	2.30680	4.4838	5.2694	0.19274	0.24597	-1
13	0.83405	3.58750	18.7777	24.5104	2.57173	3.10032	-1
14	0.06114	0.26300	1.3766	1.7968	0.18853	0.22728	-1

The output data set is in the standard OUTEST format as described for Output 2.28. The _PCOMIT_ variable specifies the number of deleted components; when it is missing, the estimates are the ordinary least squares ones (compare Output 4.1). The _TYPE_ variable provides information on what the observation contains: PARMS and SEB identify the ordinary least squares estimates and their standard errors, and IPC and IPCSEB are those for the IPC analysis.

You can see that the residual standard deviation (_ROOTMSE_) actually decreases when only one component is deleted but then increases dramatically as additional components are deleted, suggesting that the deletion of one component may provide useful results.

You can plot the coefficients against the number of deleted coefficients. Use the following statements:

```
data inpc2; set inpc;
    if _pcomit_ = . then _pcomit_ = 0;
    symbol1 v='1' i=join l=1 c=black;
    symbol2 v='2' i=join l=1 c=black;
    symbol3 v='3' i=join l=1 c=black;
    symbol4 v='4' i=join l=1 c=black;
    symbol5 v='5' i=join l=1 c=black;
    symbol6 v='6' i=join l=1 c=black;
    symbol7 v='7' i=join l=1 c=black;
proc gplot;

plot  (occup checkin hours common wings cap rooms)*_pcomit_
/ overlay;

run;
```

The data set INCPC2 is created so that the ordinary least squares coefficients are denoted by having zero components deleted. The SYMBOL statements are used so that the plotted points for each coefficient are given by their sequential order in the model. Remember that if no symbols are specified in the PLOT statement, they are implemented in numerical order. The results are shown in Output 4.15.

Output 4.15
Plot of IPC
Coefficients

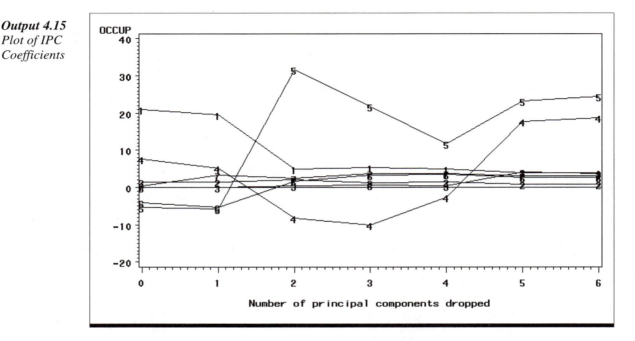

You can see that very little is changed by dropping only one component. This suggests that the resulting model will essentially be the same as the least squares one. However, there is a big change in the coefficients as two components are dropped. Remember that dropping two components also caused a dramatic increase in the residual standard deviation. Nevertheless, it may be of interest to study that model more closely. Use the following statements:

```
data incpc3; set incpc2;
   if _pcomit_ = 2 or _pcomit_= 0;
proc print data= incpc3;
   var _type_  intercept--rooms;
run;
```

The result is a printout of the coefficients and their standard errors for ordinary least squares and for omitting two components. The *t* statistics are not available and have been calculated manually[7] and inserted in the output as shown in Output 4.16.

Output 4.16
Comparing
Coefficients

Obs	_TYPE_	_RMSE_	Intercept	OCCUP	CHECKIN	HOURS	COMMON	WINGS	CAP	ROOMS
1	PARMS	284.604	171.473	21.0456	1.42632	-0.08927	7.6503	-5.3023	-4.07475	0.33191
2	SEB	284.604	148.862	4.2890	0.33071	1.16353	8.4384	9.4528	3.30195	6.81399
	T		1.152	4.9069	4.313	-0.0076	0.9066	-0.5361	-1.2340	0.4871
3	IPC	516.309	165.455	4.8342	2.12881	0.40818	-8.2067	31.6931	1.56356	2.48111
4	IPCSEB	516.309	269.677	0.6392	0.56778	2.09566	10.4837	13.5187	1.07460	0.75550
	T		0.613	7.5629	3.4973	0.1948	-0.7828	2.3444	1.4925	3.2841

---

[7] These can be computed with SAS by using PROC TRANSPOSE, which converts the observations to variables that can then be used to compute the *t* statistics.

Because this is an exploratory analysis, exact *p* values cannot be calculated from these *t* values. However, the relative magnitudes of the *t* values do provide useful information. It appears that OCCUP and CHECKIN are still important, although the magnitude of the OCCUP coefficient is smaller. However, the CAP and ROOMS coefficients now appear to have some positive effect. The apparently dramatic change in the COMMON coefficient does not, however, seem to be important. It does appear that the partial principal component regression has provided additional information on some of the factors affecting manpower requirements.

4.5.2 Ridge Regression

The least squares estimates of a standardized regression are obtained by

$$\hat{\beta} = \mathbf{R}_{xx}^{-1} \mathbf{R}_{xy},$$

where $\mathbf{R}_{xx}$ is the correlation matrix of the independent variables, and $\mathbf{R}_{xy}$ is the one-column matrix of correlations of the independent variables with the dependent variable.

The cause of multicollinearity is the existence of large correlations among the independent variables. If you now add a small constant, call it *k*, to the diagonal elements of $\mathbf{R}_{xx}$, the effective correlations have all been artificially reduced by the factor $[1/(1+k)]$. The resulting correlation matrix is

$$\mathbf{R}_{xx,k} = \mathbf{R}_{xx}(\mathbf{I} + \mathbf{D}_k)$$

where $\mathbf{D}_k$ is a diagonal matrix with diagonal values of k, which can be used to estimate a set of biased regression coefficients:

$$\hat{\beta}_k = [\mathbf{R}_{xx}(\mathbf{I} + \mathbf{D}_k)]^{-1} \mathbf{R}_{xy}.$$

This equation is called a ridge regression estimate, which is a function of the *ridge constant*, *k*.

The larger the value of *k*, the smaller the effective correlations are among the independent variables, but the larger is the bias of the estimates. The problem in ridge regression is to find some optimum compromise value for *k*.

The most popular method is to compute ridge regression estimates for a set of values of *k* starting with *k*=0 (the unbiased estimate). In many applications, a plot of the coefficients against *k* shows that as the value of *k* increases from zero, the coefficients involved in multicollinearities change rapidly. However, as *k* increases further, these coefficients change more slowly until all coefficients tend to get smaller at a rather uniform rate. Examining such a plot and picking that value of *k* where the rapid changes cease, that is, where the coefficients settle down, provides for a method of choosing *k*. This is a very subjective procedure, and although other, more objective procedures have been proposed, none has gained universal acceptance.

Using PROC REG, the MODEL option

    RIDGE = *list;*

requests a ridge regression analysis for values of the ridge constant, *k*, specified in the *list*. The *list* may consist of specific values separated by commas or may be in the form of

    RIDGE = A TO B BY C;

The additional PROC REG statement and option specification

        PLOT / RIDGEPLOT;

provide the plot of coefficients against the ridge constant. The analysis provides no additional printed output but produces an OUTEST data set identical to that produced by the PCOMIT= option. This data set can be used to obtain further information on the ridge regression results.

Use the following statements:

```
symbol1 v='1'   c=black;
symbol2 v='2'   c=black;
symbol3 v='3'   c=black;
symbol4 v='4'   c=black;
symbol5 v='5'   c=black;
symbol6 v='6'   c=black;
symbol7 v='7'   c=black;
proc reg outest =out outseb noprint;
   model manh = OCCUP CHECKIN HOURS COMMON WINGS CAP ROOMS /
   ridge= 0 to .2 by .02;
   plot /ridgeplot;
run;
```

The ridge plot is shown in Output 4.17.

Output 4.17
Ridge Plot

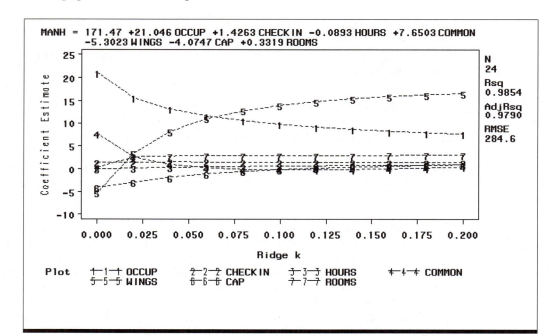

Unlike the IPC plot, there are no dramatic changes in the coefficients. This is rather typical of ridge plots. Picking a value of *k* is quite subjective, and the value 0.06 would appear to be reasonable. Use the following statements to obtain the desired results:

```
data out2; set out;
   if _RIDGE_=. or _RIDGE_ = .06;
proc print data=out2 ;
   var _type_ _RMSE_ Intercept -- rooms;
run;
```

The results are shown in Output 4,18; as in the IPC analysis, the *t* statistics have been computed manually.

Output 4.18
Comparing
Coefficients

Obs	_TYPE_	_RMSE_	Intercept	OCCUP	CHECKIN	HOURS	COMMON	WINGS	CAP	ROOMS
1	PARMS	284.604	171.473	21.0456	1.42632	-0.08927	7.65033	-5.3023	-4.07475	0.33191
2	SEB	284.604	148.862	4.2890	0.33071	1.16353	8.43835	9.4528	3.30195	6.81399
	T		1.152	4.9069	4.3130	-0.0076	0.9066	-0.5361	1.2340	0.4871
3	RIDGE	390.097	148.085	11.5852	1.53002	0.44319	0.20635	10.9902	-1.12569	2.89265
4	RIDGESEB	390.097	193.620	1.4905	0.32845	1.45768	7.05094	9.2421	0.93250	0.71136
	T		0.764	7.7727	4.6583	0.30403	0.0297	1.1891	1.2071	4.0664

The ridge regression results have some similarities to those for the IPC analysis. The coefficients for OCCUP, CHECKIN, and ROOMS appear important. However, the residual mean square is considerably smaller, and the coefficient for WINGS does not appear important. One reason for these differences is probably the importance of observation 24 in the deleted seventh component, which was significant in the (complete) principal component regression.

4.6 Summary

This chapter has presented several methods for detecting multicollinearity and three types of remedial methods for alleviating the effects of multicollinearity. It can be argued that the three types have different purposes:

❑ Model redefinition, including complete principal component regression, is used to investigate the structure of the relationships.

❑ Variable selection is used to find the smallest (most economical) set of variables needed for estimating the dependent variable.

❑ Biased estimation is used to find coefficients that may have more definite interpretations.

These different objectives are, of course, not mutually exclusive, and most regression analyses involve all of these to some degree. It is, however, useful to keep these in mind when planning a strategy for analysis.

However, because it is the easiest to use, variable selection is the most frequently used method. This is unfortunate, not only because it may be the incorrect strategy, but also because there are some negative side effects from using variable selection.

Remember that variable selection in no way guarantees that the chosen subset of variables is correct for the population. In fact, the existence of multicollinearity increases the risk of choosing the wrong subset. If variables that belong in the model have been deleted, estimates for coefficients for the remaining variables are biased. And if the selection process has failed to delete variables that do not belong, the selection process has not achieved its goal.

Another negative aspect of variable selection has already been noted; that is, p values obtained with selected models may not be taken literally. There is, in fact, no theory that specifies what the true p values are, except that they are probably quite a bit larger than those printed by the output.

If variable selection is used, it is important to stress that, if possible, the use of prior information to choose a set of suitable variables is preferable to automated variable selection. In other words, the brute force of the computer is no substitute for knowledge about the data and existing relationships. For example, the relative cost of measuring the different independent variables should have a bearing on which variables to keep in a model that you use to predict values of the dependent variable. Another case in point is the polynomial model where a natural ordering of parameters exists, which suggests a predetermined order of variable selection. See Chapter 5.

Automatic variable selection can product inferior results. An interesting demonstration of this is to randomly split a data set into two parts and perform these methods on each part. The differences in the results are often quite striking and informative. For this reason, results of analyses that are produced with variable selection should be interpreted with special caution.

Chapter 5 Curve Fitting

5.1 Introduction

In the regression models presented in earlier chapters, straight lines have described almost all relationships among variables. When relationships are not necessarily described by straight lines, there are two distinct types of analyses:

❑ curve fitting, sometimes referred to as smoothing, where the purpose of the statistical analysis is to define a relatively smooth curve or surface that describes the behavior of the response variable without any reference to a meaningful model

❑ specifying a well-fitting regression model with parameters that have physical interpretations.

This chapter presents some methods for curve fitting and their implementation with the SAS System.

The most popular regression model used for curve fitting is the *polynomial* model, in which the dependent variable is related to functions of the powers of one or more independent variables. Because such models are still linear in the parameters, PROC REG or PROC RSREG, as well as PROC GLM, can be used to implement polynomial models with the SAS System. These applications are presented in Sections 5.2, "Polynomial Models with One Independent Variable," and 5.4, "Polynomial Models with Several Variables." Because plots of the resulting response curves are usually the most important products of curve fitting, procedures for producing such plots are explained in some detail in Sections 5.3, "Polynomial Plots," 5.5, "Response Surface Plots," and 5.6, "A Three-Factor Response Surface Experiment."

A smooth curve does not necessarily have to be represented by a regression model with parameters. A frequently used curve fitting procedure that does not explicitly use parameters is the moving average. A moving average can be obtained with PROC EXPAND, which is presented in Section 5.7, "Curve Fitting without a Model." More recent developments along this line include nonparametric regression, where models without any specific parameters are fitted using least squares. One such method is provided by PROC LOESS, which is presented in Section 5.7.

Methods that attempt to fit models with meaningful parameters are described in Chapters 6, "Special Application of Linear Models," and 7, "Nonlinear Models."

5.2 Polynomial Models with One Independent Variable

A polynomial model using one independent variable is defined as follows:

$$y = \beta_0 + \beta_1 x + \beta_2 x^2 + \ldots + \beta_m x^m + \varepsilon$$

where y represents the dependent variable and x the independent variable. The highest exponent, or power, of x used in the model is known as the *degree* of the model, and it is customary for a model of degree m to include all terms with lower powers of the independent variable. The use of polynomial models is sometimes justified by the well-known Taylor series, which suggests that any continuous function can be approximated within a limited range by a polynomial equation.

Although the polynomial model describes a curvilinear response, it is still a linear regression model because it is linear in the parameters. In other words you can use the values of the required powers of the independent variable as the set of independent variables in a multiple linear regression model to provide a regression analysis for a polynomial model. Consequently, all statistics and estimates produced by the implementation of that model have the same connotation as in any linear regression analysis, although the practical implications of some of the results may differ.

Polynomial models are primarily used for curve fitting because the individual coefficients of the polynomial model itself are of little practical use, and therefore the degree of polynomial required to fit a set of data is not usually known a priori. Therefore, it is customary to build a polynomial model by sequentially fitting equations with higher-order terms until a satisfactory degree of fit has been accomplished. In other words, you start by fitting a simple linear regression of y on x. Then you specify a model with linear and quadratic terms to determine whether adding the quadratic term improves the fit by significantly reducing the residual mean square. You then continue by adding and testing the contribution of a cubic term, then a fourth-power term, and so on, until no additional terms appear to be needed.

A polynomial regression model is illustrated here using data collected to determine how growth patterns of fish are related to temperature. A curve describing how an organism grows with time is called a *growth curve*. It usually shows rapid initial growth, which gradually becomes slower and may eventually cease. Mathematical biologists have developed many sophisticated models to fit growth curves (see Section 7.3, "Fitting a Growth Curve with the NLIN Procedure"). However, a polynomial model can provide a convenient and easy approximation to such curves.

Fingerlings of a particular species of fish were put into each of four tanks that were kept at temperatures of 25, 27, 29, and 31 degrees Celsius, respectively. After 14 days, and weekly thereafter, one fish was randomly selected from each tank and its length was measured. The data from this experiment are given in Output 5.1.[1]

---

[1] To save space, a PUT statement has been used to present the data with each column representing the weights for each temperature.

Output 5.1
Data Set
FISH

TEMP	25	27	29	31
AGE				
14	620	625	590	590
21	910	820	910	910
28	1315	1215	1305	1205
35	1635	1515	1730	1605
42	2120	2110	2140	1915
49	2300	2320	2725	2035
56	2600	2805	2890	2140
63	2925	2940	3685	2520
70	3110	3255	3920	2710
77	3315	3620	4325	2870
84	3535	4015	4410	3020
91	3710	4235	4485	3025
98	3935	4315	4515	3030
105	4145	4435	4480	3025
112	4465	4495	4520	3040
119	4510	4475	4545	3177
126	4530	4535	4525	3180
133	4545	4520	4560	3180
140	4570	4600	4565	3257
147	4605	4600	4626	3166
154	4600	4600	4566	3214

The data set FISH consists of 21 observations from each tank for a total of 84 observations containing the variables AGE, TEMP, and LENGTH. The data for the fish kept at 29 degrees are used to approximate the growth curve with a fourth-degree polynomial in AGE for the dependent variable LENGTH. (The entire data set is used later.)

The polynomial regression is estimated using PROC REG, with LENGTH as the dependent variable, and the fourth-degree polynomial with AGE, AGE^2, AGE^3, and AGE^4 as independent variables.

Before continuing, it is important to point out that the estimation of a polynomial regression model may be subject to severe round-off errors. There are two reasons for this:

❑ The numerical values of powers of numbers can become very large. Mixing small and large numbers in the computation of sums of squares can create round-off errors.

❑ Powers of x are highly correlated, especially if the range of numbers is large. This creates a built-in multicollinearity, which was seen in Chapter 4 to be a potential source of round-off error.

There are two ways to counter the round-off problem:

❑ You can use the ORTHOREG procedure, which employs special methods to reduce round-off errors. However, PROC ORTHOREG does not offer any of the various diagnostics and special output options available with PROC REG and should therefore be used only if necessary. PROC ORTHOREG was used for this example and produced identical results to those from PROC REG, suggesting that round off is not a problem with this model.

❑ You can scale the independent variables. For example, scaling variables to have a range between −1 and +1 almost eliminates correlations among odd and even powers and eliminates the problem of very large numbers. Remember that scaling will not affect the fit of the model although it does affect the values of the coefficients, which are usually not important.

In order to use PROC REG for a polynomial regression, it is first necessary to generate the values of the powers of AGE to use as independent variables. This is done in the following DATA step, which in this case also selects the data for TEMP=29.[2] This data set is then used for PROC REG as follows:

```
data temp29; set fish;
   if temp = 29;
   asq = age*age;
   acub = age*age*age;
   aqt = asq*asq;
proc reg data=temp29;
   model length = age asq acub aqt / ss1 seqb;
run;
```

Note that the MODEL options SS1 and SEQB are used to obtain the Type I sums of squares and sequential parameter estimates. The results of these statements appear in Output 5.2.

Output 5.2
Polynomial
Regression
Using PROC
REG

The REG Procedure
Model: MODEL1
Dependent Variable: LENGTH

Analysis of Variance

Source	DF	Sum of Squares	Mean Square	F Value	Pr > F ①
Model	4	38250790	9562698	724.31	<.0001
Error	16	211239	13202		
Corrected Total	20	38462029			

Root MSE	114.90174	R-Square	0.9945	
Dependent Mean	3524.61905	Adj R-Sq	0.9931	
Coeff Var	3.25998			

Parameter Estimates

Variable	DF	Parameter Estimate ③	Standard Error	t Value	Pr > |t|	Type I SS ②
Intercept	1	280.18822	287.02299	0.98	0.3435	260881728
AGE	1	-2.80282	20.35328	-0.14	0.8922	30305600
ASQ	1	1.95644	0.45369	4.31	0.0005	7484973
ACUB	1	-0.02158	0.00394	-5.47	<.0001	34496
AQT	1	0.00006628	0.00001167	5.68	<.0001	425721

Sequential Parameter Estimates
④

Intercept	AGE	ASQ	ACUB	AQT
3524.619048	0	0	0	0
1143.959307	28.341187	0	0	0
-816.641969	90.969090	-0.372785	0	0
-1037.459406	103.279306	-0.545668	0.000686	0
280.188216	-2.802825	1.956435	-0.021584	0.000066279

[2] Alternately these statements may be used in the DATA step in which the data are initially read.

The circled numbers in Output 5.2 have been added to key the descriptions that follow:

① The test for the entire model is statistically significant since the *p* value for the test for MODEL is less than 0.0001. The value for the coefficient of determination (R-Square) is 0.9945 and shows that the model accounts for a large portion of the variation in fish lengths. The residual standard deviation (Root MSE) is 114.9 and indicates how well the fourth-degree polynomial curve fits the data.

② The Type I sums of squares, also called *sequential* sums of squares, are used to determine what order of polynomial model is needed. The Type I sums of squares give the contribution to the Model SS for each independent variable as it is added to the model in the order listed in the Model statement (see Section 2.4.2 "SS1 and SS2: Two Types of Sums of Squares"). Dividing these Type I sums of squares by the residual mean square provides *F* statistics that are used to test whether these additional contributions to the regression sum of squares justify addition of the corresponding terms to the model.[3]

In this example, the Type I sum of squares for the linear regression on AGE is 30,305,600; dividing by the residual mean square of 13,202 gives an *F* ratio of 2295, which clearly establishes that a linear regression fits better than no regression.[4] The additional contribution of the quadratic term, designated ASQ, is 7,484,973, and the *F* ratio is 566.94. Therefore, the addition of this term can be justified. The *F* ratio for adding the cubic term is 2.61. This is not statistically significant, so the inclusion of this term may not be justified. Because some authorities recommend continuing until two successive terms are not needed, you can continue to check for additional terms. The *F* ratio for adding the fourth-degree term is 32.25, which is significant, providing evidence that this term should be included. You could, of course, continue if you had specified a higher-order polynomial in the MODEL statement. However, polynomials beyond the fourth degree are not often used.

③ In this example, it is appropriate to recommend the fourth-degree polynomial. The estimated equation, obtained from the portion of the output labeled Parameter Estimate, is

$$\widehat{LENGTH} = 280.19 - 2.8028(AGE) + 1.9564(AGE)^2 - 0.02158(AGE)^3 + 0.00006628(AGE)^4.$$

As noted, the polynomial model is only used to approximate a curve. Therefore, the individual polynomial terms have no practical interpretation, and the remainder of the statistics for the coefficients are of little interest.[5]

④ If the tests based on the Type I sums of squares had indicated that a lower-order polynomial would suffice, the coefficients for such lower-order polynomial regression models are found under the heading Sequential Parameter Estimates. These were produced by PROC REG using the MODEL statement option SEQB. In this portion of the output, the first line is the zero-order polynomial, that is, the mean of the dependent variable (3524.62). The second line contains the coefficients of the first-order, or linear, regression:

$$1143.96 + 28.3412(AGE).$$

[3] PROC GLM allows the implementation of a polynomial model without having to generate the powers of the independent variable in the DATA step and also gives the *F* ratios for the Type I sums of squares. However, PROC GLM does not have as many desirable diagnostic and output options for regression. Therefore, it is not necessarily the best procedure to use, even if it is a bit more convenient for this purpose.
[4] The Type I sum of squares for INTERCEPT is the correction for the mean. This can be used to test the hypothesis that the mean is zero, which is seldom of interest.
[5] Remember that a partial regression coefficient is the change in the response due to a unit change in the independent variable, holding constant the other variables. Since it is impossible to change, say, x^2 holding x constant, the coefficients measure relationships that really do not exist.

The third line contains the coefficients of the quadratic model, and so on. The last line contains the coefficients of the full, or in this case, the fourth-order polynomial and is the same as the list of coefficients under Parameter Estimate in number ③.

5.3 Polynomial Plots

In testing for the appropriate degree of polynomial required to fit the growth curve, you can see that while the fourth-degree term was required, it was also evident that the major improvement in fit occurred with the addition of the quadratic term. A plot illustrating how the fit of a polynomial regression improves with the addition of higher-order terms may provide information to help decide if the improvement due to adding additional terms is indeed worthwhile.

In order to construct such a plot, you need the estimated values associated with the linear, quadratic, cubic, and fourth-order regression model estimates. This is accomplished with the following statements:

```
proc reg data = temp29;
   model length = age;
   output out=l p=pl r=rl;
   model length = age asq;
   output out=q p=pq r=rq;
   model length = age asq acub;
   output out=c p=pc r=rc;
   model length = age asq acub aqt;
   output out=qt p=pqt r=rqt;
data all; merge l q c qt;
run;
```

The four MODEL statements are for the linear, quadratic, cubic, and fourth-order polynomial models. The four OUTPUT statements create data sets with the predicted and residual values for these models. The predicted values are PL, PQ, PC, and PQT, respectively. The residual values, RL, RQ, RC, and RQT are used later. The MERGE statement places all of these into one data set named ALL.

The following statements produce the desired plot:

```
symbol1 v= star I=none c=black;
symbol2 v='1' l=1 i=spline c=black;
symbol3 v='2' l=1 i=spline c=black;
symbol4 v='3' l=1 i=spline c=black;
symbol5 v='4' l=1 i=spline c=black;
proc gplot data=all ;
plot length*age=1 pl*age=2 pq*age=3 pc*age=4 pqt*age=5 /
overlay;
run;
```

Note that the star symbol without a line joining the points is used for the data points, and the power of the polynomial is used as the plotting symbols for the estimated lines. The I=SPLINE option provides for a smooth line joining all points. The resulting plot is shown in Output 5.3.

Output 5.3
Plot of
Successive
Polynomial
Curves

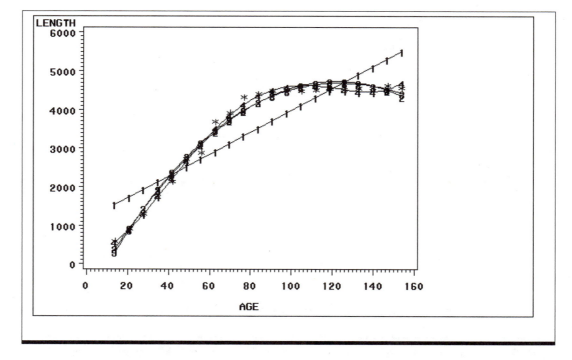

This plot clearly shows the limited improvement due to the cubic and fourth-degree terms. In fact, the fourth-order polynomial curve shows a peculiar hook at the upper end that is not typical of growth curves. Of course, the quadratic curve shows negative growth in this region, so it is also unsatisfactory. In other words, the polynomial model may be unsatisfactory for this data set. Alternate models for use with this data set are presented in Section 7.3, "Fitting a Growth Curve with the NLIN Procedure."

Another method for checking the appropriateness of a model is to plot the residual values. The following statements obtain the plot of residual values from the quadratic polynomial (for which the predicted and residual values are in data set Q):

```
proc gplot data=q;
   plot rq*age / vref=0;
run;
```

The resulting plot, which appears in Output 5.4, shows a systematic pattern that is typical of residual plots when the specified degree of polynomial is inadequate. In this case, the W-shaped pattern is due to the fourth-degree term, which was statistically significant.

Output 5.4
Residual
Plot for
Quadratic
Model

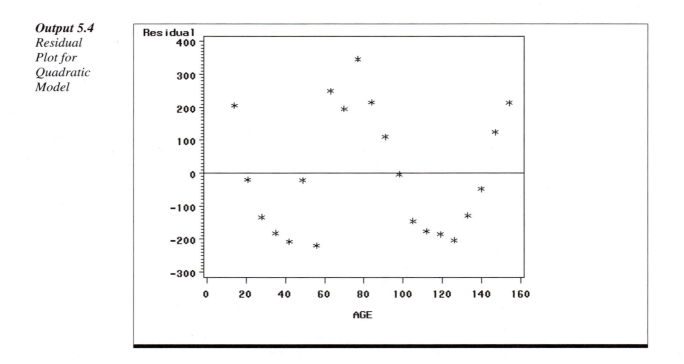

5.4 Polynomial Models with Several Variables

Polynomial models for several variables contain terms involving powers of the various variables as well as crossproducts among these variables. The crossproduct terms measure interactions among the effects of the variables. For example, consider the two-variable model:

$$y = \beta_0 + \beta_1 x_1 + \beta_2 x_2 + \beta_{12} x_1 x_2 \quad .$$

The coefficient β_{12} indicates how the linear effect of x_1 is affected by x_2, or vice versa. This is apparent in the following rearrangement of terms:

$$y = + \beta_0 + (\beta_1 + \beta_{12} x_2) x_1 + \beta_2 x_2 \quad .$$

The coefficient β_{12} specifies how the linear effect of x_1 changes with x_2. This change is linear in x_2. This effect is referred to as the linear-by-linear interaction. The interaction effect is symmetric since it also shows how the linear coefficient in x_2 is affected by x_1. Crossproducts involving higher-order terms have similar meanings, although their interpretation may be more difficult.

You can generate the desired power and crossproduct variables in the DATA step and implement the multiple regression analysis using PROC REG for such models. However, one consequence of the greater complexity of models with polynomials in several variables is that the sequential sums of square (Type I) are no longer useful in selecting the appropriate degree of model. Likewise, the partial sums of squares (Type II) are of little use since, as was the case for one-variable polynomial models, it is not customary to omit lower-order terms. A similar restriction also applies to crossproduct terms. For example, if a crossproduct of two linear terms has been included, you should include the individual linear terms. In such models, the following questions are of primary interest:

❏   Does the entire model help to explain the behavior of the response variable?

❏   Are all factors or variables needed?

❏   Is there a need for quadratic and higher-order terms?

❏   Is there a need for crossproduct terms?

❏   Is the model adequate?

Obviously, the statistics supplied by a single run of a regression model using PROC REG cannot answer all of these questions.

Since polynomial models with several variables can easily become extremely cumbersome and consequently difficult to interpret, it is common practice to restrict the degree of polynomial terms to be used for such models. The most frequently used model of this type is called the *quadratic response surface model*, in which the maximum total exponent of any term is two. In other words, this model includes all linear, quadratic, and pairwise crossproducts of linear terms in the individual variables.

The RSREG procedure (response surface regression) is the preferred SAS procedure for building and evaluating such a quadratic response surface model. The implementation of this procedure is illustrated by estimating the quadratic response surface regression relating the LENGTH of fish to AGE and TEMP using the data in Output 5.1.

The procedure is implemented with the following statements:

```
proc rsreg data=fish;
   model length = age temp;
run;
```

The MODEL statement requires only the listing of the dependent and independent variables or factors, because PROC RSREG creates the necessary squares and products variables. For this example, the regression model estimated by the procedure is

$$\widehat{\text{LENGTH}} = \beta_0 + \beta_1(\text{AGE}) + \beta_2(\text{AGE})^2 + \beta_3(\text{TEMP})$$
$$+ \beta_4(\text{TEMP})^2 + \beta_5(\text{AGE})(\text{TEMP}) \quad .$$

Options are available in this procedure to create an output data set containing predicted, residual, and other statistics associated with individual observations. These options are presented in Section 5.5, "Response Surface Plots." Another option (not illustrated) is the COVARIATES option, which allows the inclusion of independent variables that are not part of the response surface model. This option is useful for adjusting estimates for experimental conditions. The results of implementing PROC RSREG on the fish data appear in Output 5.5.

Output 5.5
Quadratic Response Surface Regression Using PROC RSREG

```
                        The RSREG Procedure

              Coding Coefficients for the Independent Variables

                 Factor    Subtracted off     Divided by

                 AGE         84.000000        70.000000
                 TEMP        28.000000         3.000000

                Response Surface for Variable LENGTH

              Response Mean                3153.345238
              Root MSE                      262.038473
              R-Square                          0.9611
              Coefficient of Variation          8.3099

                            Type I Sum
    Regression       DF     of Squares    R-Square    F Value    Pr > F

    Linear       ②    2     107867300      0.7835     785.47    <.0001
    Quadratic    ②    2      21762116      0.1581     158.47    <.0001
    Crossproduct ③    1       2682926      0.0195      39.07    <.0001
    Total Model  ①    5     132312342      0.9611     385.39    <.0001

                                    Sum of
          Residual         DF       Squares      Mean Square

     ⑤    Total Error      78       5355805         68664

                                                                  Parameter
                                                                   Estimate
                                     Standard                     from Coded
    Parameter    DF     Estimate       Error    t Value  Pr > |t|     Data

    Intercept    1        -56025   5625.614035    -9.96   <.0001   3978.557760
    AGE          1    123.822451      8.988307    13.78   <.0001   1835.337662
    TEMP         1   3934.862900    401.275363     9.81   <.0001   -297.078571
    AGE*AGE      1     -0.266708      0.017852   -14.94   <.0001  -1306.870189
    TEMP*AGE     1     -1.885584      0.301652    -6.25   <.0001   -395.972727
    TEMP*TEMP    1    -69.205357      7.147685    -9.68   <.0001   -622.848214

           ④                 Sum of
         Factor    DF        Squares    Mean Square   F Value   Pr > F

         AGE        3      121756815      40585605     591.07   <.0001
         TEMP       3       13238453       4412818      64.27   <.0001

          Canonical Analysis of Response Surface Based on Coded Data

                                     ⑥
                              Critical Value
                 Factor      Coded      Uncoded

                 AGE       0.775671    138.296998
                 TEMP     -0.485049     26.544854

            Predicted value at stationary point: 4762.416003

                              Eigenvectors
            Eigenvalues         AGE           TEMP

            -569.675554     -0.259376      0.965776
           -1360.042850      0.965776      0.259376

              Stationary point is a maximum.
```

Because the interpretation of results from the search for the optimum response (see number ⑥ below) may be affected by the scales of measurement of the factor variables, they are coded to have maximum and minimum values of +1 and -1 for the computations required for this procedure. This rescaling also reduces possible effects due to round-off error. However, all of the statistics in the output are, except where noted, converted to the original scales.

The circled numbers in Output 5.5 have been added to key the descriptions that follow:

① The *F* ratio of 385.39 indicates a statistically significant model.

② The need for linear and quadratic terms is established in this portion of the output. The line labeled Linear tests the effectiveness of the model including only the strictly linear terms AGE and TEMP. The line labeled Quadratic tests the additional contribution of the quadratic terms, $(AGE)^2$ and $(TEMP)^2$. Both linear and quadratic terms are needed, although these statistics do not specify which of these are needed.

③ The line labeled Crossproduct is a test for all crossproduct terms in the model (in this case there is only one). This term should also be included in the model.

④ The contribution of the variables, or factors, is evaluated in this portion. Here, the line labeled AGE tests the hypothesis that the factor AGE can be omitted from the model. That is, it tests the hypothesis that all terms involving the AGE factor, namely, AGE, $(AGE)^2$, and (AGE)(TEMP), may be omitted without decreasing the fit of the model. Obviously, the factor AGE should not be omitted from the model. Similarly, the line labeled TEMP determines if the factor TEMP can be omitted. Although this factor contributes less than AGE, it also appears to significantly help describe the growth of the fish.

⑤ The error sum of squares and mean square are given here. A lack-of-fit test for the adequacy of the model is provided in this part of the output if replicated observations exist. Since such observations are not available in this example, only the error mean square is printed. An example of the lack-of-fit test is provided in Section 5.6, "A Three-Factor Response Surface Experiment."

⑥ Because response surface analysis is sometimes performed to obtain information on optimum estimated response, PROC RSREG also supplies information to assist in determining if the estimated response surface exhibits such an optimum. First, the partial derivatives of the estimated response surface equation are calculated and a stationary point is found. The values of the factors for the stationary point are given in the Critical Value column, and the estimated response at this point is denoted as Predicted value at stationary point. For this example, the stationary point is AGE=138.3 and TEMP=26.5, where the estimated response is 4762.4. The program does not check to see if these values are in the range of data. The critical values are also given for the coded variables. Because the stationary point may be a maximum, minimum, or saddle point, a canonical analysis is performed to ascertain which of these it is. In this case, the point is a maximum. The eigenvalues and eigenvectors required for the analysis yield additional information on the shape of the response surface (Myers 1976). See also the RIDGE statement in Section 5.6.

Additional statistics on the output include some overall descriptive measures, the residual standard deviation (Root MSE), R-Square, and the Coefficient of Variation. Also given are the statistics for coding the factor variables. The various statistics for the individual coefficients, including the coefficients for the coded variables, are given in the columns named [Parameter] Estimate and Parameter Estimate from Coded Data. However, as noted before, the statistics for the individual terms are normally not particularly useful.

5.5 Response Surface Plots

As in the case of a one-variable polynomial regression, graphic representations are a useful device for showing the nature of the estimated response curve or surface. For multidimensional polynomial regressions, such plots are called response surface plots. A popular plot of this type is a contour plot, in which contours of equal response are plotted for a grid of values of the independent variables. A line printer contour plot may be obtained using PROC PLOT; or a high resolution one may be obtained with PROC GCONTOUR. Also available are three-dimensional representations, which may be performed using the PROC G3D.

The response surface plot for the estimated quadratic response surface for the fish lengths produced in the previous section is used to illustrate this type of plot. In addition, a residual plot may be useful for investigating specification error or detecting possible outliers. Plots involving more than two independent variables are presented in Section 5.6.

All SAS System plotting procedures construct plots that illustrate relationships among values of variables in a data set. Therefore, in order to produce a response surface plot, it is first necessary to produce a data set consisting of a grid of values of the independent variables to be used in the plot. This is not difficult since, in the SAS System, all regression procedures provide the predicted values for data points that are not in the data set used to estimate the regression model.[6]

The first step is to generate the grid of values of the independent variables. The number of grid points is somewhat arbitrary but should be sufficient to provide for a relatively smooth graphical representation. About 30 rows and columns are used for this example. The following statements produce a grid of values in the F1 data set:

```
data f1;
   do temp = 25 to 31 by .2;
      do age = 10 to 160 by 5;
         flag = 1;
output; end; end;
   run;
```

The data set F1 has 31*31=961 observations, containing the variables TEMP and AGE, and a variable FLAG, whose value is unity and is used later in this section. The next step is to concatenate the data set F1 with the original data set (FISH):

```
data f2;
   set fish f1;
   run;
```

Note that the data set FISH does not have the variable FLAG, and F1 does not have the variable LENGTH, so these variables are denoted as missing when they do not exist in data set F2. If you now run PROC RSREG on data set F2, the estimation is based only on the 72 observations (from data set FISH) for which the dependent variable is not missing. However, predicted values, when requested, are computed for all observations.

---

[6] Such predicted values can also be produced with the OUTEST option and SCORE procedure which, however, require more programming for this application.

The instructions for producing a data set with estimated values and other statistics as well as the format of the resulting output are different for PROC RSREG than for PROC REG or PROC GLM. The following statements are required:

```
proc rsreg data=f2 out=p1;
   model length = age temp /predict residual;
   id flag;
run;
```

The PROC RSREG statement must include the option OUT=P1, which specifies that the output data set containing the predicted values is data set P1. The MODEL statement options PREDICT and RESIDUAL request that the output data set contains the predicted and residual values. The ID FLAG option is required so that the ID variable FLAG is included in the output data set. Other statistics, such as confidence intervals and the PRESS and Cook's D statistics, may be obtained by adding other keywords. For more information about the RSREG procedure, see the *SAS/STAT User's Guide*.

The output data set produced by PROC RSREG has a different format than the output data set produced by PROC REG. To illustrate this format, the first 20 observations from the data set produced from the instructions given above are reproduced in Output 5.6.

Output 5.6
Partial
Listing of
Output Data
Set

Obs	flag	AGE	TEMP	_TYPE_	LENGTH
1	.	14	25	PREDICT	114.61
2	.	14	25	RESIDUAL	505.39
3	.	21	25	PREDICT	586.04
4	.	21	25	RESIDUAL	323.96
5	.	28	25	PREDICT	1031.34
6	.	28	25	RESIDUAL	283.66
7	.	35	25	PREDICT	1450.50
8	.	35	25	RESIDUAL	184.50
9	.	42	25	PREDICT	1843.53
10	.	42	25	RESIDUAL	276.47
11	.	49	25	PREDICT	2210.42
12	.	49	25	RESIDUAL	89.58
13	.	56	25	PREDICT	2551.16
14	.	56	25	RESIDUAL	48.84
15	.	63	25	PREDICT	2865.78
16	.	63	25	RESIDUAL	59.22
17	.	70	25	PREDICT	3154.25
18	.	70	25	RESIDUAL	-44.25
19	.	77	25	PREDICT	3416.59
20	.	77	25	RESIDUAL	-101.59

The data set P1 has one observation for each statistic requested in the MODEL statement and contains the following variables:

❑ the independent variables or factors (AGE and TEMP).

❑ a variable called _TYPE_, which identifies the output statistic. In this example, two statistics, PREDICT and RESIDUAL, were requested; hence two observations are created for each observation in the input data set. One observation, identified by _TYPE_=PREDICT, has the predicted values, and the second observation, identified by _TYPE_=RESIDUAL, contains the residual values. The predicted and residual values are identified by the name of the response variable(s).

❑ the variables used in a BY or ID statement or COVAR option, which is the variable FLAG.

The format for the data set P1 is not directly suitable for each of the desired plots. This is because the data set for the residual plot must contain the predicted and residual values for the originally observed data, while the data set for the contour plot requires only the values of the factor variables and predicted values for the grid data set. The data sets are created as follows:

```
data plot
    pred (rename=(length=pred))
    resid (rename=(length=resid));
set p1;
    if flag = 1 and _type_ = 'PREDICT' then output plot;
        if flag = . and _type_ = 'PREDICT' then output pred;
        if flag = . and _type_ = 'RESIDUAL' then output resid;
    end;
data resid2;
    merge pred resid;
run;
```

The data set PLOT is the grid data set P1 in which FLAG=1. The predicted values, which the OUTPUT statement gives the name of the response variable, are renamed PRED.

The residual plot requires a data set that contains both the predicted and residual values for the observations of the original data set FISH for which FLAG= . (missing). To obtain this data set, the first DATA step creates two data sets for which FLAG is missing: the data set named PRED containing the predicted values, and the data set named RESID containing the residual values.[7] Finally, a second DATA step produces data set RESID2 by merging PRED and RESID. The residual plot is obtained as follows:

```
proc plot data = resid2;
    plot resid*pred / vref = 0;
run;
```

The plot appears in Output 5.7. Although there appear to be some patterns in these residuals, they are not sufficiently consistent to identify specification errors.

Output 5.7
Residual
Plot

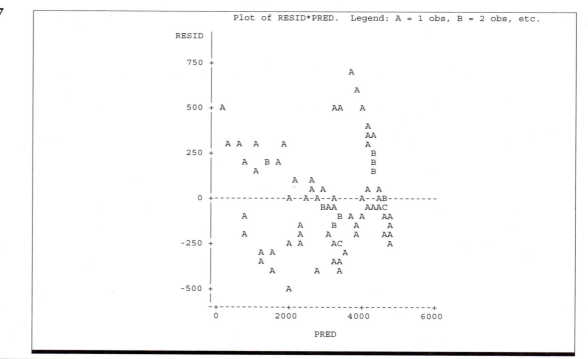

---

[7] It is actually easier to produce these data sets with PROC REG, which has a format for the output data set that is easier to use, but then you must remember to create the polynomial variables in both the input and grid data sets.

Use the following statements to produce the desired contour plot:

```
proc gcontour data=plot;
plot temp*age=length/ llevels=1 3 20 33 35 36 40 clevels=black
black black black black black black;
run;
```

The PLOT statement specifies TEMP as the row variable and AGE as the column variable. The contours represent seven levels (the default number of levels) of the estimated response (LENGTH), with values specified in the legend at the bottom of the plot. The CLEVELS and LLEVELS options are needed to provide different types of black lines, as the default output uses colors to identify the different lines. The resulting plot appears as Output 5.8A. The plot shows the growth of fish with age, although the predicted growth becomes negative with increasing age. The plot also shows that fish grow faster at temperatures of 26 to 28 degrees Celsius.

Output 5.8A
Contour Plot

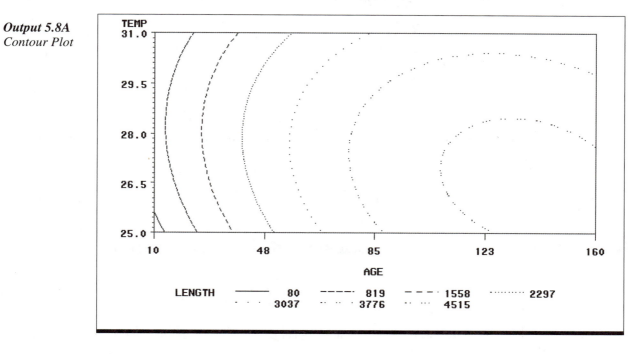

Alternately, you can produce a three-dimensional graph of this surface using PROC G3D. The required statements are

```
proc g3d;
plot temp*age=length;
```

The results are shown in Output 5.8B.

Output 5.8B
Three-Dimensional Response Surface Plot

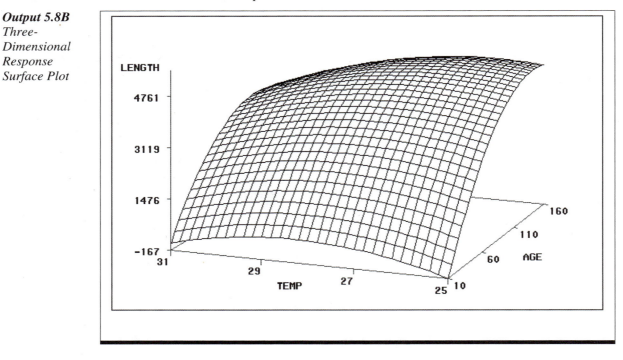

The three-dimensional plot gives a better picture of the shape of the response curve, but it is not as informative as to the values of the estimated responses.

5.6 A Three-Factor Response Surface Experiment

The data for this example come from an experiment concerning a device for automatically shelling peanuts. Dickens and Mason (1962) maintain that

> . . . peanuts flow through stationary sheller bars and rest on the grid which has perforations just large enough to pass shelled kernels. The grid is reciprocated and the resulting forces on the peanuts between the moving grid and the stationary bars break open the hulls. . . . the problem became one of determining the combination of bar grid spacing, length of stroke and frequency of stroke which would produce the most satisfactory performance. The performance criteria are . . . kernel damage, shelling time and unshelled peanuts.

The paper cited above describes three separate experiments; the second one is used for this illustration. The experimental design is a three-factor composite design consisting of 15 points, with 5 additional observations at the center point (Myers 1976). The data consist of responses resulting from the shelling of 1000 grams of peanuts. The factors of the experiment are the following:

LENGTH	length of stroke (inches)
FREQ	frequency of stroke (strokes per minute)
SPACE	bar grid spacing (inches).

The response variables are the following:

TIME	time needed to shell 1000 grams of peanuts (minutes)
UNSHL	unshelled peanuts (grams)
DAMG	damaged peanuts (percent).

Output 5.9 shows the peanut sheller data.

Output 5.9
Peanut
Sheller Data
for Response
Surface
Analysis

Obs	LENGTH	FREQ	SPACE	TIME	UNSHL	DAMG
1	1.00	175	0.86	16.00	284	3.55
2	1.25	130	0.63	9.25	149	8.23
3	1.25	130	1.09	18.00	240	3.15
4	1.25	220	0.63	4.75	155	5.26
5	1.25	220	1.09	15.50	197	4.23
6	1.75	100	0.86	13.00	154	3.54
7	1.75	175	0.48	3.50	100	8.16
8	1.75	175	0.86	7.00	176	3.27
9	1.75	175	0.86	6.25	177	4.38
10	1.75	175	0.86	6.50	212	3.26
11	1.75	175	0.86	6.50	200	3.57
12	1.75	175	0.86	6.50	160	4.65
13	1.75	175	0.86	6.50	176	4.02
14	1.75	175	1.23	12.00	195	3.80
15	1.75	250	0.86	5.00	126	4.05
16	2.25	130	0.63	4.00	84	9.02
17	2.25	130	1.09	7.00	145	3.00
18	2.25	220	0.63	2.25	97	7.41
19	2.25	220	1.09	5.75	168	3.78
20	2.50	175	0.86	3.50	168	3.72

Because this experiment has six replications at the center point (observations 8 through 13, where LENGTH=1.75, FREQ=175, and SPACE=0.86), you can obtain an estimate of pure error and consequently perform a test for lack of fit (Freund and Wilson 1998, Section 6.3). For this example, only the variable UNSHL is analyzed. You may want to perform analyses for the other responses and ponder the problem of multiple responses. Use the following SAS statements:

```
proc rsreg data=peanuts;
   model unshl = length freq space / lackfit;
   ridge min;
run;
```

The LACKFIT option in the MODEL statement computes the pure error sum of squares from all observations occurring within identical factor level combinations. In this example, the statistic comes from the six replications identified above.

This sum of squares is subtracted from the residual sum of squares (from the model) to obtain the lack-of-fit sum of squares. This quantity indicates the additional variation that can be explained by adding to the model all additional parameters allowed by the construct of the treatment design. Thus, the ratio of the resulting lack of fit and pure error mean squares provides a test for the possible existence of such additional model terms. In other words, it is a test for the adequacy of the model.

The RIDGE statement is explained below in the section on the canonical analysis. The results of the RSREG procedure appear in Output 5.10. The circled numbers in Output 5.10 have been added to key the descriptions of items that are different from those for Output 5.5.

Output 5.10
PROC
RSREG with
LACKFIT
Option

```
                        The RSREG Procedure

            Coding Coefficients for the Independent Variables

             Factor     Subtracted off      Divided by

             LENGTH         1.750000         0.750000
             FREQ         175.000000        75.000000
             SPACE          0.855000         0.375000


               Response Surface for Variable UNSHL

            Response Mean              168.150000
            Root MSE                    19.222839
            R-Square                     0.9144
            Coefficient of Variation    11.4320

                                Type I Sum
         Regression      DF       of Squares    R-Square   F Value   Pr > F

         Linear           3            27642     0.6401     24.94    <.0001
         Quadratic        3            10989     0.2545      9.91    0.0024
         Cross-product    3       856.375000     0.0198      0.77    0.5354
         Total Model      9            39487     0.9144     11.87    0.0003


                  ①                   Sum of
         Residual        DF          Squares   Mean Square  F Value   Pr > F

         Lack of Fit      5      1903.675307   380.735061     1.06    0.4742
         Pure Error       5      1791.500000   358.300000
         Total Error     10      3695.175307   369.517531

                                                                    Parameter
                                                                    Estimate
                                           Standard                 from Coded
         Parameter    DF      Estimate        Error    t Value  Pr > |t|   Data

         Intercept     1    -130.386632   210.102546    -0.62    0.5488  184.350375
         LENGTH        1    -319.784414   111.964075    -2.86    0.0171  -50.515924
         FREQ          1       3.006709     1.180000     2.55    0.0290   -5.684025
         SPACE         1     789.775578   232.083714     3.40    0.0067   52.104286
         LENGTH*LENGTH 1      52.110564    23.973433     2.17    0.0548   29.312192
         FREQ*LENGTH   1       0.405556     0.302058     1.34    0.2091   22.812500
         FREQ*FREQ     1      -0.009684     0.002524    -3.84    0.0033  -54.474119
         SPACE*LENGTH  1      -1.086957    59.098259    -0.02    0.9857   -0.305707
         SPACE*FREQ    1      -0.471014     0.656647    -0.72    0.4896  -13.247283
         SPACE*SPACE   1    -331.287196   100.067743    -3.31    0.0079  -46.587262

                             Sum of
         Factor    DF       Squares   Mean Square  F Value   Pr > F

         LENGTH     4         16591   4147.864308    11.23    0.0010
         FREQ       4   6462.244483   1615.561121     4.37    0.0266
         SPACE      4         17546   4386.601785    11.87    0.0008

                                                              Continued
```

Output 5.10
(Continued)
PROC
RSREG with
LACKFIT
Option

```
                                    ②
                 Canonical Analysis of Response Surface Based on Coded Data

                                Critical Value
                  Factor        Coded         Uncoded

                  LENGTH       0.842171        2.381628
                  FREQ         0.057503      179.312738
                  SPACE        0.548273        1.060602

              Predicted value at stationary point: 177.199118

                                          Eigenvectors
          Eigenvalues          LENGTH           FREQ             SPACE

            30.850993         0.990957         0.133509        -0.013376
           -43.243358         0.072745        -0.450813         0.889649
           -59.356824        -0.112747         0.882578         0.456448

                  Stationary point is a saddle point.

                  Estimated Ridge of Minimum Response for Variable UNSHL

      Coded        Estimated        Standard              Uncoded Factor Values
      Radius        Response          Error        LENGTH         FREQ          SPACE

       0.0        184.350375        7.789077       1.750000     175.000000      0.855000
       0.1        176.949575        7.766784       1.796364     175.374746      0.825583
       0.2        169.149158        7.714192       1.831772     175.248503      0.792137
       0.3        160.752551        7.676618       1.858325     174.550897      0.756422
       0.4        151.620194        7.730364       1.878628     173.258186      0.719767
       0.5        141.657267        7.977460       1.894789     171.369838      0.682995
       0.6        130.797679        8.525890       1.908265     168.898882      0.646596
       0.7        118.992786        9.459711       1.920022     165.870412      0.610878
       0.8        106.204777       10.818957       1.930682     162.321155      0.576038
       0.9         92.403027       12.603536       1.940650     158.297515      0.542198
       1.0         77.562104       14.791639       1.950185     153.852091      0.509420
```

The portions of Output 5.10, which are similar to those of Output 5.5, show that

❑ The overall model is significant. The linear and quadratic terms are needed, but the crossproduct terms are not, indicating that there are no interactions. That means that the response curve for any one factor has a similar shape across levels of the other factors.

❑ The factors LENGTH and SPACE are needed; the factor FREQ is marginally significant with a *p* value of 0.027.

❑ The Lack of Fit test (item ①) shows that the Total Error, the residual sum of squares from the nine-term model, is 3695.1753, with 10 degrees of freedom. The Pure Error, the sum of squares among the six replicated values, is 1791.5 with 5 degrees of freedom. The difference, 1903.6753, with 5 degrees of freedom, is the additional sum of squares that could be obtained by adding five terms to the model. The *F* statistic derived from the ratio of the lack-of-fit to pure error mean square has a *p* value of 0.4742. You may conclude that additional terms are not needed.

❑ The canonical analysis (Item ②) shows that the response surface has a saddle point. This means it has no point at which the response is either maximum or minimum.

Since you are looking for a minimum amount of unshelled peanuts, the existence of a saddle point may be disappointing. Of course it is possible that the saddle point is not well defined and that a broad range of points may provide a guide for finding optimum operating conditions.

The RIDGE MIN statement provides help in determining where optimum response may occur by providing sets of values of the factor levels producing the fastest decrease in the estimated response starting at the stationary point. The results are in the last section of Output 5.10 and show that increasing LENGTH but decreasing FREQ and SPACE lowers the estimated percent of unshelled peanuts. The statement RIDGE MAX will look for the direction of fastest increase in the estimated response.

It is not possible to produce a three-factor response surface plot, but you can produce a plot that represents the response curve for two factors for several levels of the third factor. Since there are three different factor combinations for such plots, you must choose which combination is most useful. Often this decision is not easy to make and several possibilities may have to be explored. In this example, the factor FREQ has the smallest effect, so it appears logical to plot the response to LENGTH and SPACE for selected values of FREQ.

The response surface plot is produced here for LENGTH and SPACE for FREQ values of 175, 200, and 225. The residual plot is also obtained.

As before, you must first generate the data representing the grid of points needed for the response surface plot. Use the following statements:

```
data p1;
   flag=1;
      do freq = 175 to 225 by 25;
         do length = 1 to 2.5 by .05;
            do space = .6 to 1.2 by .02;
               output;
            end;
         end;
      end;
run;
```

The data set P1 consists of a 31-by-31 grid of values of LENGTH and SPACE for three values of FREQ. The data set P1 is concatenated with the original data set and PROC RSREG is used as follows:

```
data p2;
   set peanuts p1;
proc rsreg data=p2 out=p3;
   model unshl = length freq space / predict residual;
   id flag;
run;
```

Next, you create the data sets for the plots using the same procedure that generated the data for Outputs 5.7 and 5.8:

```
data plot (rename = (unshl = predict))
     p4 (rename = (unshl = predict))
     p5 (rename = (unshl = residual));
set p3;
   if flag = 1 and _type_ = 'PREDICT' then output plot;
   if flag = . and _type_ = 'PREDICT' then output p4;
   if flag = . and _type_ = 'RESIDUAL' then output p5;
data p6;
   merge p4 p5;
run;
```

The data set PLOT contains the predicted values for the grid needed for the response surface plots. The data sets P4 and P5 contain the predicted and residual values for the actual data points, since the variable FLAG was undefined in that set of data. Finally, sets P4 and P5 are merged to produce the data set P6 required for the residual plot.

The residual plot is constructed with the following statements:

```
proc gplot data = p6;
    plot residual*predict / vref = 0;
```

Output 5.11 shows the residual plot that is produced by the preceding statements.

Output 5.11
Residual Plot

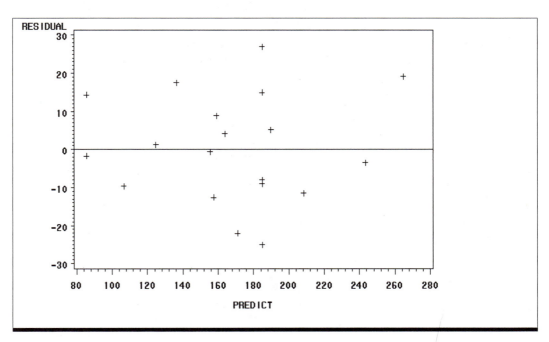

No obvious outliers or specification appear in this plot, confirming the results of the lack-of-fit analysis.

Either PROC GCONTOUR, PROC PLOT, or PROC G3D may be used to produce response surface plots. For PROC G3D, these statements are needed:

```
proc g3d data=plot;
plot length*space=predict/zmin=80 zmax=280 zticknum=3;
by freq.
run;
```

In order to compare the plots for the three values of FREQ it is useful to have the same scales for these plots. In most SAS plotting procedures this is accomplished by AXIS specifications. However, for PROC G3D, this is accomplished by specifying the minimum and maximum values and the number of tick marks for each axis. The axes are referenced by x, y, and z corresponding to the list of variables in the plot statement of the form PLOT $x*y=z$. In this case the actual x and y values are the same for each plot; hence it is only necessary to specify the scale of the z axis corresponding to UNSHL. Output 5.12 shows the output from PROC G3D.[8]

---

[8] The PROC GREPLAY statements required to place these three on one plot are not shown.

Output 5.12
Response
Surface Plots

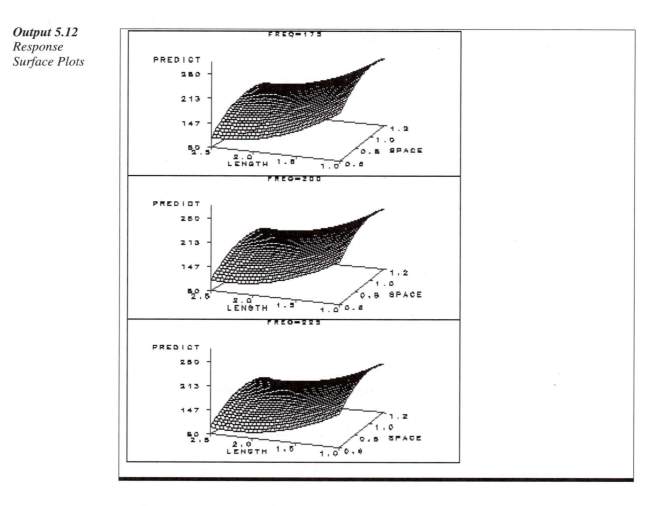

From these plots you can see that

❑   the shape of the response surface is almost identical for all values of FREQ, confirming the lack of interaction (crossproducts) terms. The percent of unshelled peanuts changes very little with FREQ but appears to be higher for FREQ=200.

❑   the saddle point is not of interest.

❑   the percentage of unshelled peanuts decreases with increasing LENGTH and decreasing SPACE. These results confirm the results from the RIDGE statement.

Further experiments for finding optimum operating conditions for unshelled peanuts should concentrate in the middle regions of FREQ, higher regions of LENGTH, and lower regions of SPACE. However, this recommendation is subject to the behavior of the other response variables.

5.7 Curve Fitting without a Model

Although polynomial regression uses a specified algebraic model to fit a curve, the function itself is of little interest. Curve fitting without the use of a specific model, also called smoothing, has been done for years, especially with time series. The simplest smoothing method is to simply draw a freehand line that gives a reasonably smooth approximation to the data. Two methods that are less subjective are presented in this section:

❑ the moving average. This is a popular method that has been used for many years, especially in smoothing time series data. However, the method has no objective criteria for evaluating the fit of the curve.

❑ nonparametric curve fitting using PROC LOESS. Because this method uses least squares, some reasonably objective criteria for evaluating the fit are available.

5.7.1 The Moving Average

The moving average approximates or estimates each point of the curve with the mean of a number of adjacent observations. The number of observations to be used is arbitrary: a smaller number gives a better fit while a larger number gives a smoother curve. For example, values for a five-point centered moving average are obtained as

$$MA_t = (y_{t-2} + y_{t-1} + y_t + y_{t+1} + y_{t+2})/5,$$

with special adjustments for the first and last two observations of the series.[9] Moving averages can be computed with the SAS System using PROC CONVERT available in SAS/ETS software.

The moving average is illustrated with data on total exports (EXPORT) from the nation of Barbados (Barbados dollars) for the years (YEAR) 1967–1993. The data for the data set BARBEXP are shown in Output 5.13, which also includes the five-point moving averages (EXP_MOV) as computed by PROC CONVERT (see below).

Output 5.13
Barbados
Exports

Obs	year	export	exp_mov
1	1967	53.518	56.841
2	1968	59.649	58.158
3	1969	57.357	57.162
4	1970	62.106	59.079
5	1971	53.182	63.890
6	1972	63.103	77.529
7	1973	83.700	100.752
8	1974	125.555	117.643
9	1975	178.218	135.233
10	1976	137.638	155.783
11	1977	151.055	177.209
12	1978	186.450	209.024
13	1979	232.684	240.897
14	1980	337.291	285.211
15	1981	297.004	349.954
16	1982	372.627	420.151
17	1983	510.165	451.987
18	1984	583.668	476.709
19	1985	496.471	445.086
20	1986	420.614	392.659
21	1987	214.511	325.995
22	1988	248.029	275.665
23	1989	250.350	239.826
24	1990	244.820	251.201
25	1991	241.420	256.043
26	1992	271.384	257.467
27	1993	272.242	261.682

[9] There is no universally adopted procedure for obtaining the end points. The SAS procedure used here simply uses the available observations.

Use the following statements to produce the moving averages:

```
proc expand data = barbexp out=barbexp2;
convert export = exp_mov / transform =(cmovave 5);
```

The CONVERT statement with the CMOVAVE =5 option produces a five-point, centered moving average of the specified variable (EXPORT) to be named EXP_MOV. The procedure produces no printed output and the results are placed in the data set specified by the PROC option OUT= BARPEXP2. The PRINT procedure provides the results shown in Output 5.13.

The resulting smoothed curve, shown in Output 5.14, is produced by the following statements:

```
symbol1 c=black v=star;
symbol2 c=black v=point l=1 i=join;
proc gplot data=barbexp2;
plot export*year=1 exp_mov*year=2 / overlay ;
```

Output 5.14
Plot of
Moving
Average

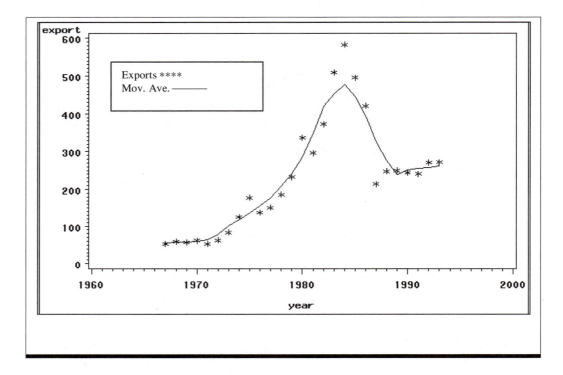

The moving average appears to provide a relatively smooth curve. As noted, using more points in the moving average will produce a smoother curve, which may hide certain features. Using fewer points will produce a more precise approximation with a less-smooth curve.

5.7.2 Nonparametric Smoothing with PROC LOESS

The availability of computing power has enabled the development of somewhat more formal procedures for smoothing, whose results are more readily evaluated by statistical methods. PROC LOESS provides one such procedure. This procedure uses weighted least squares to fit linear or quadratic functions in neighborhoods, with the weights being a smooth decreasing function of the distance from the center of each neighborhood. The final response curve is a blending of these individual curves. The number of neighborhoods dictates the smoothness of the resulting response: the fewer the neighborhoods, the smoother but less well fitting is the curve. The number of neighborhoods is specified by the *smoothing parameter*, which specifies the proportion of the total number of observations in each neighborhood. Thus the maximum smoothing parameter of 1 essentially fits a linear or quadratic function for the entire data set, which is obviously the smoothest yet poorest fitting curve, while small values, such as 0.1 may actually fit the curve to all data points. References for this method are provided in the *SAS/STAT User's Guide*.

The following set of statements implements PROC LOESS on the Barbados export data using a number of values of the smoothing parameter and creates a data set with predicted values:

```
proc loess data=barbexp;
model export=year /smooth = .2 .3 .4 .5;
ods output outputstatistics=a;
```

❏ The model statement specifies the independent and dependent variable(s). The MODEL option SMOOTH = specifies that the model will be fit with smoothing parameters 0.2, 0.3, 0.4, and 0.5, respectively. This option may also be stated in the form

SMOOTH = .2 TO.5 BY .1;

❏ The ODS OUTPUT statement specifies the various data sets to be created.[10] The keyword OUTPUTSTATISTICS specifies that the output data set A is to be created. This data set contains coordinates and fit results for the data points. The default output, which is used here, includes only the predicted values, which are found as the variable DEPVAR. Other variables in the data set include the dependent and independent variables and the smoothing parameter named SMOOTH.

The output from PROC LOESS is shown for only one smoothing parameter (0.2) in Output 5.15.

Output 5.15
Output from
PROC
LOESS

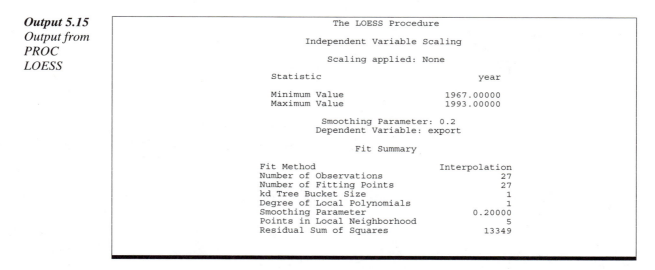

```
                      The LOESS Procedure

                  Independent Variable Scaling

                     Scaling applied: None

        Statistic                          year

        Minimum Value                 1967.00000
        Maximum Value                 1993.00000

              Smoothing Parameter: 0.2
              Dependent Variable: export

                        Fit Summary

        Fit Method                    Interpolation
        Number of Observations                  27
        Number of Fitting Points                27
        kd Tree Bucket Size                      1
        Degree of Local Polynomials              1
        Smoothing Parameter                0.20000
        Points in Local Neighborhood             5
        Residual Sum of Squares              13349
```

Most of the output contains technical information on the fitting process. For example, it shows that the default linear function (degree of polynomial = 1) is used in each neighborhood. Most of these

---

[10] As previously noted, using ODS output provides a more flexible method for creating output data.

are of little interest at this point. The statistic of primary interest is the residual sum of squares, which may be useful to see how the fit changes with the different values of the smoothing parameter. These values are

Smoothing Parameter	Residual SS
0.5	95,211
0.4	68,964
0.3	44,109
0.2	13,329

You can see the residual sums of squares decrease quite rapidly as the smoothing parameter decreases. However, the best way to judge the effectiveness of the procedure is to plot the response curves. The following statements produce the desired plots:[11]

```
symbol1 c=black v=dot;
symbol2 c=black v=point i=join l=1;
proc gplot data=a;
by smoothingparameter;
plot depvar*n=1 pred*n=2 / overlay;
```

The resulting plots are shown in Output 5.16.

Output 5.16
PROC
LOESS
Curves

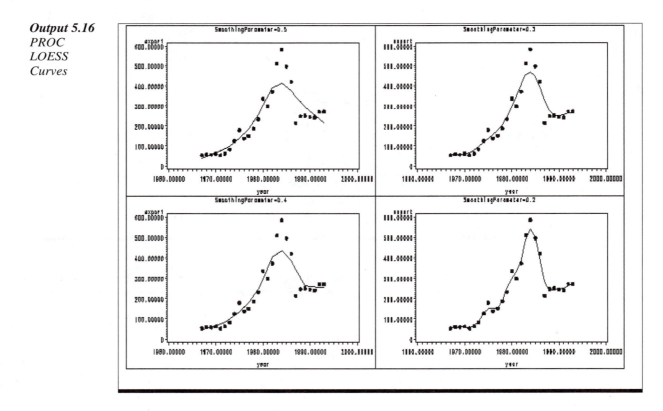

---

[11] Again the PROC GREPLAY statements are not shown.

The decision on which of these is best is quite subjective. Obviously using a smoothing parameter of 0.2 provides the best fit, but the line may not be sufficiently smooth to be considered representative of trends.

PROC LOESS has a number of additional options. These include

❑ PROC LOESS can perform smoothing for models with several dependent and independent variables.

❑ Residuals from the predicted values are available to be included in the output data set.

❑ Because actual regressions are performed in the neighborhoods, approximate degrees of freedom for the model and error are available. These are in turn used to compute a residual mean square, which is, in turn, used to provide confidence intervals, and *t* statistics (studentized residuals), which can be output to a data set for plotting or other purposes.

❑ The various results printed in the output from the procedure are also available in a data set that is produced as an ODS OUTPUT with the keyword FITSUMMARY.

❑ A number of options are available for fine-tuning the estimation method. These may be especially useful for more complex models.

Finally, nonparametric smoothing can also be performed with the SAS System by PROC TSPLINE, which uses a different estimation method.

You may want to compare the results of nonparametric smoothing with the results obtained by fitting a polynomial to the Barbados export data. You can obtain a fifth-degree polynomial, which is illustrated here with the use of PROC GLM, requesting that only the Type I sums of squares be printed. Note also that the variable N, which is the observation number created in the DATA step, is used instead of YEAR to reduce possible round-off error.

```
proc glm data=barbexp;
model export = n n*n n*n*n n*n*n*n n*n*n*n*n/ ss1;
```

The results are shown in Output 5.17.

Output 5.17
Fifth-Degree
Polynomial

```
                              The GLM Procedure

                    Number of observations      27
                        The GLM Procedure

Dependent Variable: export

                                    Sum of
    Source                 DF        Squares      Mean Square    F Value    Pr > F

    Model                   5     493869.0066     98773.8013      23.80    <.0001
    Error                  21      87152.7275      4150.1299
    Corrected Total        26     581021.7341

                R-Square     Coeff Var      Root MSE      export Mean

                0.850001     28.03277      64.42150       229.8078

    Source                 DF      Type I SS      Mean Square    F Value    Pr > F

    n                       1     245800.0475     245800.0475     59.23    <.0001
    n*n                     1     104506.5295     104506.5295     25.18    <.0001
    n*n*n                   1      64566.8464      64566.8464     15.56    0.0007
    n*n*n*n                 1      25719.2284      25719.2284      6.20    0.0213
    n*n*n*n*n               1      53276.3548      53276.3548     12.84    0.0018

                                    Standard
         Parameter      Estimate        Error     t Value    Pr > |t|

         Intercept    -58.9493529    106.6688061    -0.55     0.5863
         n            118.4555164     71.9569597     1.65     0.1146
         n*n          -33.9837264     15.2209339    -2.23     0.0366
         n*n*n          3.8995927      1.3498561     2.89     0.0088
         n*n*n*n       -0.1757176      0.0527561    -3.33     0.0032
         n*n*n*n*n      0.0026881      0.0007503     3.58     0.0018
```

It can be seen that the fifth-order polynomial model will be needed, although the greatest improvements occur with the third-degree function. Note that the residual sum of squares of the fifth-degree polynomial compares with that from PROC LOESS with a smoothing parameter between 0.3 and 0.4.

You can use the following statements, which employ PROC REG, to successively fit the polynomials, place the predicted values into data sets, and produce plots of the second- to fifth-degree polynomials. The resulting plots are shown in Output 5.18.

```
proc reg data=barbexp noprint;
    model export= n n2;
    output out=a p=p2;
    model export=n n2 n3;
    output out=b p=p3;
    model export = n n2 n3 n4 ;
    output out=c p=p4;
    model export = n n2 n3 n4 n5;
    output out=d p=p5;
    symbol1 c=black v=star l=none;
    symbol2 c=black v=point l=1 i=join;
data e; merge a b c d;
proc gplot data=e ;
    goptions nodisplay;
    plot export*n=1 p2*n=2 / overlay;
    plot export*n=1 p3*n=2 / overlay;
    plot export*n=1 p4*n=2 / overlay;
    plot export*n=1 p5*n=2 / overlay;
run;quit;
```

Output 5.18
Successive
Polynomial
Fits

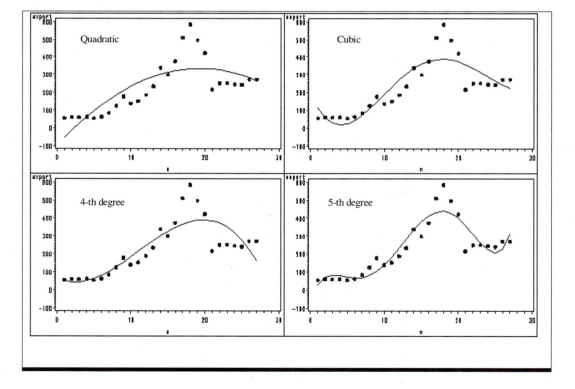

The polynomial fits are indeed much smoother than those produced by PROC LOESS. However, you can see that the polynomial model cannot fit some of the more irregular features and may thus be considered too smooth.

5.8 Summary

Curve fitting, sometimes referred to as smoothing, consists of methods that attempt to fit a smooth curve or surface to a set of data points. Polynomial models are most frequently used for this purpose. In the SAS System, they can be analyzed with PROC REG and PROC RSREG, although PROC GLM can also be useful since it allows the direct use of polynomial terms.

Because a traditional model with parameters is not really required for curve fitting, so called nonparametric regressions can be used for this purpose. A traditional favorite is the use of moving averages, which can be obtained with PROC EXPAND. More recently developed computer intensive methods are available with PROC LOESS, which is illustrated in this chapter, and also with PROC TPSPLINE.

An important limitation of the methods described in this chapter is that they provide only approximations to a curve, and the model coefficients either do not exist or usually have no practical interpretation. Therefore, if you want to model a response curve with a regression where the coefficients have practical interpretation, you may need to look for other types of models. Some of these are explored in Chapter 6, "Special Applications of Linear Models," and Chapter 7, "Nonlinear Models."

Chapter **6** Special Applications of Linear Models

6.1 Introduction

In Chapter 5, "Curve Fitting," you are introduced to the use of regression models whose primary purpose is to fit a relatively smooth curve or surface relating a response to some independent variables. Although these models are useful for providing visual descriptions, they do not provide for adequate interpretable relationships. For example, in the data on fish growth, the polynomial model provides a statistically significant fit but provides little information on any functional form of the relationship. Furthermore, the plot of the resulting curve, even for the fourth-order polynomial, showed some features that did not have the desired characteristics for such a curve. A number of other regression methods are available for such cases.

Methods presented in this chapter may be analyzed with the SAS System by adaptations and options of the standard linear regression procedures available in the REG procedure. Additional models can be analyzed by the LOGISTIC procedure. Chapter 7, "Nonlinear Models," introduces the estimation of models that cannot be analyzed in this manner.

Models presented in this chapter include

❑   linear regression models using logarithms of variables, which provide for estimating a multiplicative model

❑   spline functions, which allow different functions, usually polynomial models, to be fitted for different regions of the data

❑   indicator variables, which provide for the estimation of the effect of binary categorical variables

❑   models with a binary response variable

❑   linear regression models for a function of the response variable, which are useful if the response variable has some known distribution.

6.2 Multiplicative Models

A linear regression model using the logarithms of the variables is equivalent to estimating a multiplicative model. [1] The following linear model using logarithms:

$$\log(y) = \beta_0 + \beta_1\left(\log(x_1)\right) + \beta_2\left(\log(x_2)\right) + \ldots + \beta_m\left(\log(x_m)\right) + \varepsilon$$

is equivalent to the model

$$y = \left(e^{\beta_0}\right)\left(x_1^{\beta_1}\right)\left(x_2^{\beta_2}\right)\ldots\left(x_m^{\beta_m}\right)\left(e^{\varepsilon}\right) \quad .$$

In this multiplicative model the coefficients, called *elasticities,* measure the *percent* change in the dependent variable associated with a *one percent* change in the corresponding independent variable, holding constant all other variables. This means that these coefficients are not affected by the scales of measurement of the variables in the model. The intercept is a scaling factor. The error component in this model is also multiplicative and exhibits variation that is proportional to the magnitude of the dependent variable (Freund and Wilson 1998, Section 8.2).

The multiplicative model is illustrated with an example in which the weight of lumber is to be estimated from external measurements of the trees. The data set PINES contains information on a sample of individual pine trees. The dependent variable WEIGHT represents the weight of harvested lumber from a tree. The four independent variables are

HEIGHT   height of the tree

DBH   diameter of the tree at breast height (about four feet)

AGE   age of the tree

GRAV   measure of the specific gravity of the tree.

Obviously, the first two variables are relatively easy to measure; hence a model for predicting tree weights using these variables could provide a low-cost estimate of timber yield. The other two variables are included to see if their addition to the model provides for better prediction. Output 6.1 shows the data for estimating tree weights.

---

[1] Either base 10 or base *e* may be used with identical results except for a change in the definition of the intercept. Base *e* is used in all examples.

Output 6.1
Data for
Estimating
Tree Weights

Obs	DBH	HEIGHT	AGE	GRAV	WEIGHT
1	5.7	34	10	0.409	174
2	8.1	68	17	0.501	745
3	8.3	70	17	0.445	814
4	7.0	54	17	0.442	408
5	6.2	37	12	0.353	226
6	11.4	79	27	0.429	1675
7	11.6	70	26	0.497	1491
8	4.5	37	12	0.380	121
9	3.5	32	15	0.420	58
10	6.2	45	15	0.449	278
11	5.7	48	20	0.471	220
12	6.0	57	20	0.447	342
13	5.6	40	20	0.439	209
14	4.0	44	27	0.394	84
15	6.7	52	21	0.422	313
16	4.0	38	27	0.496	60
17	12.1	74	27	0.476	1692
18	4.5	37	12	0.382	74
19	8.6	60	23	0.502	515
20	9.3	63	18	0.458	766
21	6.5	57	18	0.474	345
22	5.6	46	12	0.413	210
23	4.3	41	12	0.382	100
24	4.5	42	12	0.457	122
25	7.7	64	19	0.478	539
26	8.8	70	22	0.496	815
27	5.0	53	23	0.485	194
28	5.4	61	23	0.488	280
29	6.0	56	23	0.435	296
30	7.4	52	14	0.474	462
31	5.6	48	19	0.441	200
32	5.5	50	19	0.506	229
33	4.3	50	19	0.410	125
34	4.2	31	10	0.412	84
35	3.7	27	10	0.418	70
36	6.1	39	10	0.470	224
37	3.9	35	19	0.426	99
38	5.2	48	13	0.436	200
39	5.6	47	13	0.472	214
40	7.8	69	13	0.470	712
41	6.1	49	13	0.464	297
42	6.1	44	13	0.450	238
43	4.0	34	13	0.424	89
44	4.0	38	13	0.407	76
45	8.0	61	13	0.508	614
46	5.2	47	13	0.432	194
47	3.7	33	13	0.389	66

As an initial step you can use a linear regression model and make provisions for a residual plot. The required SAS statements are

```
proc reg data=pines;
   model weight = height dbh age grav;
   plot r.*p./ vref=0;
run;
```

The results are shown in Output 6.2.

*Output 6.2
PROC REG
Output for
the Linear
Model*

```
                             The REG Procedure
                               Model: MODEL1
                          Dependent Variable: WEIGHT

                             Analysis of Variance

                                    Sum of          Mean
        Source           DF        Squares        Square    F Value    Pr > F

        Model             4        6570095       1642524     124.06    <.0001
        Error            42         556076         13240
        Corrected Total  46        7126171

                 Root MSE              115.06478    R-Square    0.9220
                 Dependent Mean        369.34043    Adj R-Sq    0.9145
                 Coeff Var              31.15413

                             Parameter Estimates

                         Parameter      Standard
        Variable    DF     Estimate        Error     t Value    Pr > |t|

        Intercept    1    -379.24822    206.69095      -1.83      0.0736
        DBH          1     170.22033     16.23793      10.48      <.0001
        HEIGHT       1       1.90016      3.01674       0.63      0.5322
        AGE          1       8.14583      4.02036       2.03      0.0491
        GRAV         1   -1192.86848    548.92715      -2.17      0.0355
```

The model does fit rather well, and the coefficients have the expected signs. Surprisingly, however, the HEIGHT coefficient is not statistically significant. Furthermore, the significance of both the AGE and GRAV coefficients $(\alpha < 0.05)$ indicates that the model using only the two-dimensional measurements is inadequate. Finally, the residual plot (see Output 6.3) shows a pattern that suggests a poorly specified model.

*Output 6.3
Residual
Plot, Linear
Model*

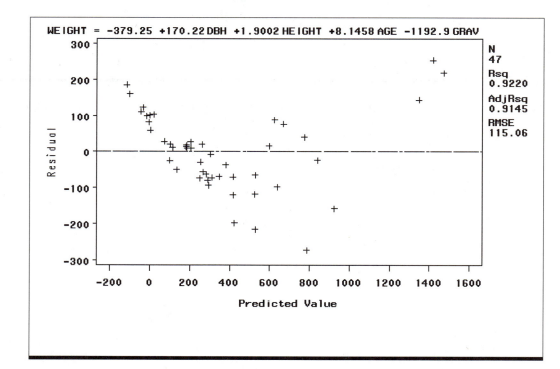

The pattern indicates the need for a curvilinear form such as, for example, a quadratic response function, as well as the possibility of heterogeneous variances. A standard deviation proportional to the mean response is a condition that suggests the use of the logarithmic transformation. Furthermore, in this case, the multiplicative model has theoretical justification. The amount of lumber in a tree is a function of the volume of the trunk, which is in the shape of a cylinder. The volume of a cylinder is $\pi r^2 h$, where r is the radius and h the height of the cylinder. Thus, a multiplicative model for volume or weight using radius (or diameter) and height is appropriate. Furthermore, it is also reasonable to expect that age and gravity have a relative or multiplicative effect. As noted, the multiplicative model of the form

$$y = e^{\beta_0} x_1^{\beta_1} x_2^{\beta_2} \ldots x_m^{\beta_m} e^{\varepsilon}$$

is converted to a linear model by taking logarithms of both sides:

$$\log(y) = \beta_0 + \beta_1 \log(x_1) + \beta_2 \log(x_2) + \ldots + \beta_m \log(x_m) + \varepsilon.$$

Because this is now a linear regression model using the logarithms of the values of the variables, it can be estimated by any procedure that can perform linear regressions. The model is implemented by obtaining the logarithms of the observed values in a DATA step:

```
data log; set pines;
   array x dbh -- weight;
   array l ldbh lheight lage lgrav lweight;
do over x;
   l = log(x);
end;
run;
```

The new variables, LDBH through LWEIGHT, are now available for use in the multiplicative model. You can use these variables with PROC REG and make provisions for the residual plot by creating a data set with PLW and RLW being the predicted and residual values for the multiplicative model as follows:

```
proc reg data=log;
   model lweight = ldbh lheight lage lgrav;
   plot rlw*plw / vref = 0;
run;
```

Output 6.4 shows the output from PROC REG.

Output 6.4
PROC REG
Output for
Multiplicative
Model

```
                            The REG Procedure
                             Model: MODEL1
                       Dependent Variable: LWEIGHT

                            Analysis of Variance

                                  Sum of        Mean
        Source           DF       Squares      Square    F Value    Pr > F

        Model             4      36.58763     9.14691     572.86    <.0001
        Error            42       0.67062     0.01597
        Corrected Total  46      37.25825


                 Root MSE            0.12636    R-Square     0.9820
                 Dependent Mean      5.49466    Adj R-Sq     0.9803
                 Coeff Var           2.29971


                          Parameter Estimates

                          Parameter      Standard
        Variable    DF     Estimate        Error    t Value   Pr > |t|

        Intercept    1     -1.55823       0.56527     -2.76     0.0086
        LDBH         1      2.14478       0.11911     18.01     <.0001
        LHEIGHT      1      0.97785       0.16961      5.77     <.0001
        LAGE         1     -0.15509       0.08050     -1.93     0.0608
        LGRAV        1      0.10775       0.26747      0.40     0.6891
```

Note that both R^2 and F statistics indicate that the overall fit of this model is much better than that of the linear model. Furthermore, both the LDBH and LHEIGHT coefficients are highly significant, while neither of the other variables contributes significantly. Finally, the coefficients for LDBH and LHEIGHT are quite close to 2.0 and 1.0, respectively, which are the values you would expect with a model based on the cylindrical shape of the tree trunk. You can, in fact test the hypothesis that these values are correct as follows:

```
dbh :test ldbh=2;

height: test lheight=1;
```

These statements provide the results in Output 6.4A, which shows that these values are consistent with the results.

Output 6.4A
Test of
Hypothesis

```
                          The REG Procedure
                          Model: MODEL1

          Test DBH Results for Dependent Variable LWEIGHT

                                     Mean
          Source          DF        Square      F Value     Pr > F

          Numerator        1        0.02359       1.48       0.2309
          Denominator     42        0.01597


         Test HEIGHT Results for Dependent Variable LWEIGHT

                                     Mean
          Source          DF        Square      F Value     Pr > F

          Numerator        1        0.00027242    0.02       0.8967
          Denominator     42        0.01597
```

The residuals from this model (see Output 6.5) appear to show virtually no patterns that suggest specification errors. There are some suspiciously large residuals for some of the smaller values of the predicted values that may require further scrutiny. Of course, these residuals show *relative* errors and therefore do not necessarily correspond to large absolute residuals. You can use additional procedures for examining residuals as presented in Chapter 3, "Observations."

Output 6.5
Residual
Plot, Log
Model

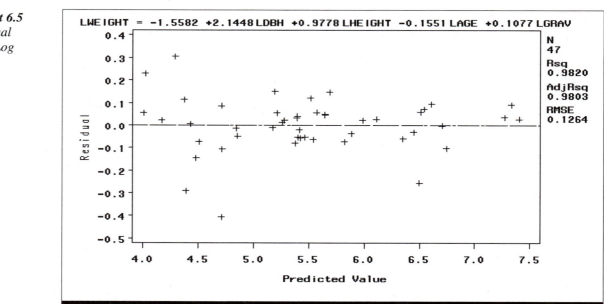

It may be of interest to see how well the multiplicative model estimates actual weights rather than logarithms of weights. You can do this by taking antilogs of the predicted values from the multiplicative model. This is done in a new DATA step:

```
data resid; set b;
   pmw = exp(plw);
   rmw = weight - pmw;
```

The variable PMW is the antilog of the predicted logarithm weights obtained by use of the EXP function, and RMW is the resulting residuals. Using the MEANS procedure to obtain the mean and sum of squares of these residuals, you obtain the following results:

mean residual = 3.970
sum of squares = 77,001.

You can see that the sum of residuals is not zero. This reflects the well-known fact that estimated or predicted values obtained in this manner are biased. However, the residual sum of squares is smaller than that obtained by the linear model. You can use PROC GPLOT to get the plot for these residuals:

```
proc gplot;
plot rmw*pmw / vref=0;
```

The results are shown in Output 6.6.

Output 6.6
Residual Plot,
Exponentiated
Log Residuals

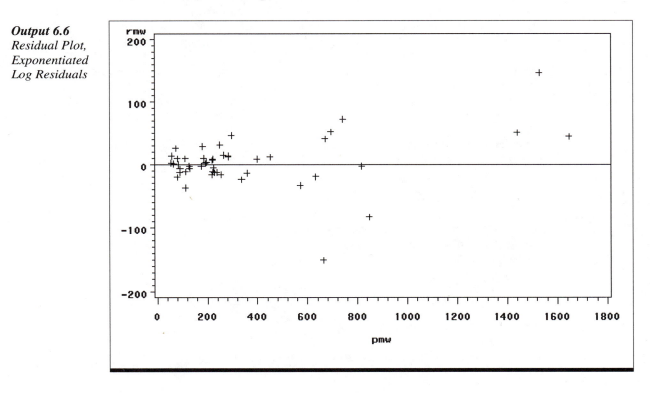

These residuals do not have the obvious specification error pattern exhibited by the residuals of the linear model. They do, however, show the typical pattern of multiplicative errors, where larger residuals are associated with larger values of the response variable.

6.3 Spline Models

In a review paper on spline models, Smith (1979) gives the following definition:

> Splines are generally defined to be piecewise polynomials of degree *n* whose function values and first (*n*−1) derivatives agree at points where they join. The abscissas of these joint points are called knots. Polynomials may be considered a special case of splines with no knots, and piecewise (sometimes also called grafted or segmented) polynomials with fewer than the maximum number of continuity restrictions may also be considered splines. The number and degrees of polynomial pieces and the number and position of knots may vary in different situations.

First consider splines with known knots, that is, splines for which the values of the independent variable are known for the joint points. Fitting spline models with known knots is much easier than with unknown knots, because with known knots you can use linear regression methods. Estimation of spline models with unknown knots requires the use of nonlinear methods, which are described in Chapter 7. The following example uses the fish data presented in Chapter 5.

You can fit a spline of two straight lines with a known knot to the fish growth at 29 degrees, as presented in Output 5.1. Judging from the plot in Output 5.3, the rate of increase is roughly constant until about AGE = 80, at which point growth appears to stop abruptly. A linear spline with a knot at AGE = 80 would appear to be suitable for these data.

In order to perform a regression analysis, you need to write a linear regression equation to represent the spline model. Define a variable, AGEPLUS, as follows:

```
data spline; set temp29;
   ageplus = max (age - 80, 0);
run;
```

The MAX function returns the maximum of the two values specified in the argument, that is, AGEPLUS = 0 when AGE < 80 and AGEPLUS = (AGE – 80) when AGE ≥ 80. Next, represent the spline model with the regression equation

$$\text{LENGTH} = \beta_0 + \beta_1 * \text{AGE} + \beta_2 * \text{AGEPLUS} + \varepsilon$$

which defines the relationship

$$\text{LENGTH} = \beta_0 + \beta_1 * \text{AGE} + \varepsilon$$

for AGE < 80, and

$$\text{LENGTH} = \beta_0 + \beta_1 * \text{AGE} + \beta_2 (\text{AGE} - 80) + \varepsilon$$

or, equivalently,

$$\text{LENGTH} = (\beta_0 - 80 * \beta_2) + (\beta_1 + \beta_2) \text{AGE} + \varepsilon$$

for AGE > 80. Notice that both expressions give the same estimated LENGTH when AGE = 80. In other words, the two line segments are joined at AGE = 80. The SAS statements for performing the regression are

```
proc reg data=spline;
   model length = age ageplus;
run;
```

Results appear in Output 6.7.

Output 6.7
Linear
Spline
Regression

```
                           The REG Procedure
                             Model: MODEL1
                        Dependent Variable: LENGTH

                           Analysis of Variance

                                   Sum of          Mean
      Source              DF       Squares        Square    F Value    Pr > F

      Model                2      38350273      19175136    3088.44    <.0001
      Error               18        111756    6208.68907
      Corrected Total     20      38462029


              Root MSE              78.79524    R-Square     0.9971
              Dependent Mean      3524.61905    Adj R-Sq     0.9968
              Coeff Var              2.23557


                           Parameter Estimates

                           Parameter     Standard
      Variable       DF     Estimate        Error    t Value    Pr > |t|

      Intercept       1    -327.61814     55.92915      -5.86    <.0001
      AGE             1      60.17531      0.97298      61.85    <.0001
      ageplus         1     -58.86310      1.63527     -36.00    <.0001
```

The fitted equation is

$$\widehat{\text{LENGTH}} = -327.62 + 60.18\ \text{AGE} - 58.86\ \text{AGEPLUS} \quad .$$

Therefore the fitted spline model for AGE < 80 is

$$\widehat{\text{LENGTH}} = -327.62 + 60.18\ \text{AGE}$$

and for AGE > 80 is

$$\widehat{\text{LENGTH}} = (-327.62 + 58.86*80) + (60.18 - 58.86)*\text{AGE}$$
$$= 4376.38 + 1.32*\text{AGE}.$$

Notice the small slope of 1.32 for AGE > 80.

You can compare the fit of the linear spline to the fit of the fourth-degree polynomial used in Chapter 5. In Output 6.7, you see MS(Error) = 6208 for the linear spline, while in Output 5.2 you see MS(Error) = 13,202 for the fourth-degree polynomial, thus indicating a better fit for the spline function.

It is often useful to fit splines that have additional conditions imposed on their parameters. For example, you may want to impose the condition on the fish growth spline that it is flat beyond AGE = 80, that is, restrict the slope to be zero for AGE > 80. From the general expression for the model for AGE > 80, you see that the slope is $\beta_1 + \beta_2$. Thus, you would want to restrict $\beta_1 + \beta_2 = 0$. You can do this in the REG procedure with the following statements:

```
proc reg data=spline;
   model length = age ageplus;
   restrict age + ageplus = 0;
run;
```

Output 6.8
Linear
Spline
Regression
with Zero
Slope after
80

```
                              The REG Procedure
                                Model: MODEL1
                          Dependent Variable: LENGTH

NOTE: Restrictions have been applied to parameter estimates.


                               Analysis of Variance

                                        Sum of        Mean
      Source              DF           Squares       Square   F Value   Pr > F

      Model                1          38335595     38335595   5760.94   <.0001
      Error               19            126434   6654.40471
      Corrected Total     20          38462029


                 Root MSE             81.57453   R-Square    0.9967
                 Dependent Mean     3524.61905   Adj R-Sq    0.9965
                 Coeff Var            2.31442


                              Parameter Estimates

                            Parameter     Standard
      Variable       DF      Estimate        Error   t Value   Pr > |t|

      Intercept       1    -358.01877     54.16288     -6.61    <.0001
      AGE             1      61.07520      0.80467     75.90    <.0001
      ageplus         1     -61.07520      0.80467    -75.90    <.0001
      RESTRICT       -1         11185   7531.35083      1.49    0.1416*

            * Probability computed using beta distribution.
```

Results in Output 6.8 produce the fitted model

$$\widehat{\text{LENGTH}} = -358.02 + 61.08*\text{AGE} - 61.08*\text{AGEPLUS} \quad .$$

For AGE < 80, you get

$$\widehat{\text{LENGTH}} = -358.02 + 61.08*\text{AGE}$$

and for AGE > 80, you get

$$\widehat{\text{LENGTH}} = (-358.02 + 61.08*80) = 4528.38.$$

This restricted spline model naturally has a larger SS(Error) than the unrestricted model (compare MS(Error) = 6209 in Output 6.7 with MS(Error) = 6654 in Output 6.8), but the difference is not very large because the slope for the unrestricted model is quite close to zero. The test in Output 6.8 labeled RESTRICT under the variable list has a *p* value of 0.7188, so the model with flat response past AGE=80 seems appropriate.[2]

---

[2] Alternately you can test AGE + AGEPLUS = 0, which will give almost the identical *p* value.

Linear spline models are sometimes criticized because the abrupt change in trend going from one segment to the next does not represent what would naturally occur. The true change in trend should be smooth. In mathematical terms, this means the fitted function should have a continuous derivative at each value of the independent variable. This is not possible for linear splines, but it is possible for quadratic splines that are joined segments of parabolas.

The general equation for a quadratic spline function for the fish growth data is

$$LENGTH = \beta_0 + \beta_1 * AGE + \beta_2 * AGEPLUS + \beta_3 * AGE2 + \beta_4 * AGEPL2 + \varepsilon$$

where AGEPLUS is as defined in the linear spline, AGE2 is the square of AGE, and AGEPL2 is the square of AGEPLUS. Using these definitions, the equation for AGE < 80 is

$$LENGTH = \beta_0 + \beta_1 * AGE + \beta_3 * AGE2 + \varepsilon$$

and for AGE > 80 the equation is

$$LENGTH = \beta_0 + \beta_1 * AGE + \beta_2 * (AGE - 80) + \beta_3 * AGE2 +$$
$$\beta_4 * (AGE - 80)^2 + \varepsilon$$

Collecting terms, this becomes

$$LENGTH = (\beta_0 - 80*\beta_2 + 80^2*\beta_4) + (\beta_1 + \beta_2 - 160*\beta_4) * AGE +$$
$$(\beta_3 + \beta_4) * AGE^2 + \varepsilon \quad .$$

Checking for the continuity condition, the derivative of the function with respect to AGE (ignoring ε) for AGE < 80 is

$$\beta_1 + 2\beta_3 * AGE2$$

and the derivative for AGE > 80 is

$$(\beta_1 + \beta_2 - 160\beta_4) + 2(\beta_3 + \beta_4) * AGE \quad .$$

Requiring the derivative to be continuous (in other words, the quadratic spline function must be smooth) means the two derivative expressions must be equal at the knot (AGE = 80). Setting them equal gives

$$(\beta_1 + \beta_2 - 160\beta_4) + 2*80(\beta_3 + \beta_4) = \beta_1 + 2*80\beta_3$$

which requires that $\beta_2 = 0$.

To fit the quadratic spline model to the fish growth data, you must first create the SAS variables AGE2 = AGE*AGE and AGEPL2 = AGEPLUS*AGEPLUS in a DATA step and perform the regression:

```
data quad; set spline;
   age2 = age*age;
   agepl2 = ageplus*ageplus;
proc reg data=quad;
   model length = age age2 ageplus agepl2 / p;
   restrict ageplus=0;
run;
```

In this example, the RESTRICT statement produces results equivalent to leaving AGEPLUS out of the model. The advantage of using the RESTRICT statement is that you get a test for the effect of the restriction.

Output 6.9
Quadratic
Spline with
Continuous
Derivative

```
                             The REG Procedure
                              Model: MODEL1
                         Dependent Variable: LENGTH

NOTE: Restrictions have been applied to parameter estimates.

                            Analysis of Variance

                                  Sum of         Mean
   Source              DF        Squares       Square   F Value   Pr > F

   Model                3       37822814     12607605    335.30   <.0001
   Error               17         639215        37601
   Corrected Total     20       38462029


            Root MSE            193.90948    R-Square     0.9834
            Dependent Mean     3524.61905    Adj R-Sq     0.9804
            Coeff Var             5.50157


                           Parameter Estimates

                      Parameter     Standard
   Variable     DF     Estimate        Error    t Value   Pr > |t|

   Intercept     1    -987.15875    249.08299     -3.96     0.0010
   AGE           1      99.00888      9.80226     10.10     <.0001
   Age2          1      -0.44551      0.08287     -5.38     <.0001
   ageplus       1    -3.185E-15            0     -Infty    <.0001
   agepl2        1       0.13197      0.14252      0.93     0.3674
   RESTRICT     -1   -8617.75220   2296.85527     -3.75     <.0001*

          * Probability computed using beta distribution.
```

Results in Output 6.9 show MS(Error) = 37601, a considerable increase from 6208 for the linear spline. This result is supported by the fact that the test for the restriction used to ensure the continuous derivative has a *p* value of less than 0.0001. Thus the quadratic spline, with conditions imposed to make the derivative continuous, is not a viable model.

The reason for the poorer fit is revealed by a residual plot, which you can get by adding the following interactive statement:

```
plot r.*age/ vref=0;
run;
```

Output 6.10
Residual Plot
for Spline
with
Continuous
Derivatives

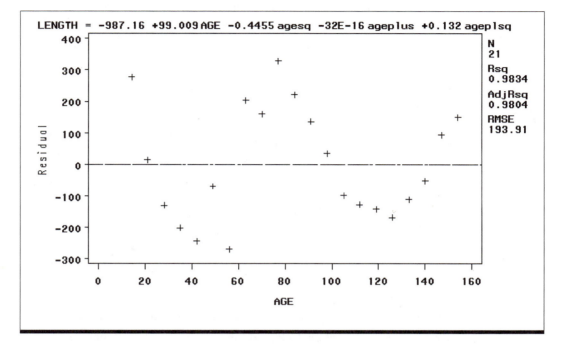

Output 6.10 shows a definite cycling of residuals, which is due to the requirement of a smooth transition of growth between the two segments, a condition that is not supported by the data.

6.4 Indicator Variables

The independent variables in a regression model are usually quantitative; that is, they have a defined scale of measurement. Occasionally, you may need to include qualitative or categorical variables in such a model. This can be accomplished by the use of indicator or dummy variables.

The example in Chapter 2 concerned a model for determining operating costs of airlines. It has been argued that long-haul airlines have lower operating costs than do short-haul airlines. Recall the variable TYPE shown in Output 2.1. This is an indicator (sometimes called dummy) variable that classifies airlines into two groups or classes. TYPE = 0 defines the short-haul lines with average stage length of less than 1200 miles and TYPE = 1 defines the long-haul lines that have average stage lengths of 1200 miles or longer.[3] Use PROC REG with the following statements:

```
proc reg;
    model cpm = utl spa alf type;
run;
```

---

[3] This is an arbitrary definition used for this example.

When TYPE = 0, INTERCEPT is the intercept for the model describing the short-haul lines. When TYPE = 1, that is, for the long-haul lines, the intercept is the sum of the INTERCEPT and TYPE coefficients. The other coefficients are the same for both classes; hence, the TYPE coefficient simply estimates the difference in levels of operating costs between the two types of airlines regardless of the values of the other variables. In other words, you have estimated a model describing two parallel planes. The results appear in Output 6.11.

Output 6.11
Using an
Indicator
Variable for
an Intercept
Shift

```
                              The REG Procedure
                                Model: MODEL1
                          Dependent Variable: CPM

                             Analysis of Variance

                                     Sum of        Mean
   Source                 DF         Squares       Square    F Value   Pr > F

   Model                   4         6.04854      1.51214      8.68    0.0001
   Error                  28         4.88013      0.17429
   Corrected Total        32        10.92867


                Root MSE              0.41748    R-Square     0.5535
                Dependent Mean        3.10570    Adj R-Sq     0.4897
                Coeff Var            13.44243


                             Parameter Estimates

                         Parameter      Standard
   Variable      DF       Estimate         Error    t Value   Pr > |t|

   Intercept      1        7.74863       0.86313       8.98    <.0001
   UTL            1       -0.13997       0.05621      -2.49    0.0190
   SPA            1       -3.55384       1.12221      -3.17    0.0037
   ALF            1       -6.23622       1.32043      -4.72    <.0001
   TYPE           1        0.01673       0.18131       0.09    0.9271
```

The positive coefficient for TYPE indicates higher costs for the long-haul lines. However, it is not statistically significant; hence, there is insufficient evidence of a difference in operating costs between the two types of airlines. The other coefficients are interpreted as before and have, in fact, quite similar values to those of the original model. This is to be expected with the insignificant TYPE coefficient.

The inclusion of the indicator variable only estimates a difference in the average operating cost, but it does not address the possibility that the relationships of cost to the various operating factors differ between types. Such a difference can be estimated by adding to the model variables that are products of the indicator and continuous variables.

A simple example illustrates this principle. Assume the following model:

$$y = \beta_0 + \beta_1 x_1 + \beta_2 x_2 + \beta_3 x_1 x_2$$

where x_1 is a continuous variable, and x_2 is an indicator variable with values 0 and 1 identifying two classes.

Then, for the first class ($x_2 = 0$), the model equation is

$$y = \beta_0 + \beta_1 x_1$$

while for the other class ($x_2 = 1$), the model equation is

$$y = (\beta_0 + \beta_2) + (\beta_1 + \beta_3)x_1 \quad .$$

In other words, the slope of the regression of y on x_1 is β_1 for the first class and $(\beta_1 + \beta_3)$ for the second class. Thus β_3 is the difference in the regression coefficient between the two classes. The coefficient β_3 is sometimes referred to as a slope shift coefficient, as it allows the slope of the regression line to shift from one class to the next.

You can use this principle to estimate different regression coefficients for the two types of airlines for the three cost factors. You need to create three additional variables in a DATA step and then perform the regression:

```
data prod; set air;
    utltp = utl*type;
    spatp = spa*type;
    alftp = alf*type;
proc reg;
    model cpm = utl spa alf type utltp spatp alftp / vif;
    alldiff : test utltp,spatp,alftp;
run;
```

Three features of the model are of interest:

❑ The TYPE variable must be included. Leaving this variable out is equivalent to imposing the arbitrary requirement that the intercept is the same for models describing both types of airlines.

❑ The variance inflation factors are not necessary for the analysis, but they do show an important feature of the results of this type of model.

❑ The TEST statement is used to test the null hypothesis that the three product coefficients are zero; that is, the effects of all three factors are the same for both types.

The results appear in Output 6.12.

Output 6.12
Using an
Indicator
Variable for
Slope Shift

```
                        The REG Procedure
                         Model: MODEL1
                     Dependent Variable: CPM

                      Analysis of Variance

                                Sum of        Mean
    Source              DF      Squares       Square    F Value    Pr > F

    Model                7      8.35794       1.19399     11.61    <.0001
    Error               25      2.57074       0.10283
    Corrected Total     32     10.92867

              Root MSE             0.32067    R-Square     0.7648
              Dependent Mean       3.10570    Adj R-Sq     0.6989
              Coeff Var           10.32523

                      Parameter Estimates

                     Parameter     Standard                        Variance
    Variable    DF    Estimate       Error    t Value   Pr > |t|   Inflation

    Intercept    1    10.64436      1.50431      7.08    <.0001             0
    UTL          1    -0.41589      0.07548     -5.51    <.0001       3.72250
    SPA          1    -6.05484      4.32444     -1.40    0.1738      46.82544
    ALF          1    -6.96499      2.09709     -3.32    0.0028       6.13091
    TYPE         1    -4.07580      1.72380     -2.36    0.0261     232.92861
    utltp        1     0.41751      0.09213      4.53    0.0001      59.18561
    spatp        1     2.27294      4.41725      0.51    0.6114     110.06340
    alftp        1     0.61188      2.41188      0.25    0.8018      95.46335

          Test ALLDIFF Results for Dependent Variable CPM

                                  Mean
    Source              DF       Square    F Value    Pr > F

    Numerator            3      0.76980       7.49     0.0010
    Denominator         25      0.10283
```

The overall model statistics show that the residual mean square has decreased from 0.174 to 0.103 and the R-Square value has increased from 0.5535 to 0.7648. In other words, allowing different coefficients for the two types of airlines has clearly improved the fit of the model. The test for the three additional coefficients given at the bottom of the output ($p = 0.0010$) indicates that the slope shifts as a whole significantly improve the fit of the model.

The statistics for the coefficients for the product variables show that among these only UTLTP is statistically significant ($p < 0.05$). In other words, the coefficient for the utilization factor is the only one that can be shown to differ between the types.

In checking the statistics for the other coefficients, note that neither SPA nor its change or shift coefficient (SPATP) is statistically significant ($p = 0.1738$ and 0.6114, respectively). This appears to contradict the results for the previous models, where the coefficient for SPA does indeed appear to be needed in the model. This apparent contradiction arises from the fact that in this type of model, just as in polynomial models (see Chapter 5), it is not legitimate to test for lower-order terms in the presence of higher-order terms. In other words, if UTLTP is in the model, the test for UTL is not meaningful. One reason for this is the high degree of multicollinearity often found in models of this type. In this example, the multicollinearity is made worse by a large difference in the sizes of planes (SPA) used by the two types of airlines. Note that the variance inflation factors involving SPA and SPATP exceed 100 (see Chapter 4, "Multicollinearity: Detection and Remedial Measures").

For this reason, it is useful to reestimate the model omitting the two insignificant slope shift terms, STATP and ALFTP. The SAS statements are

```
proc reg data=prod;
    model cpm = utl spa alf type utltp;
    utltype1: test utl + utltp;
run;
```

Note that again the TYPE variable is kept in the model since it is a lower-order term in the model that still includes one product term. The TEST statement is included to ascertain the significance of the effect of UTL for the TYPE = 1 airlines. The results appear in Output 6.13.

Output 6.13
Final
Equation for
Slope Shift

```
                          The REG Procedure
                            Model: MODEL1
                      Dependent Variable: CPM

                        Analysis of Variance

                                  Sum of        Mean
    Source            DF         Squares       Square    F Value    Pr > F

    Model              5         8.33067      1.66613      17.32    <.0001
    Error             27         2.59800      0.09622
    Corrected Total   32        10.92867

              Root MSE              0.31020    R-Square     0.7623
              Dependent Mean        3.10570    Adj R-Sq     0.7183
              Coeff Var             9.98801

                        Parameter Estimates

                     Parameter     Standard                          Variance
    Variable   DF     Estimate       Error    t Value   Pr > |t|    Inflation

    Intercept   1     10.14142      0.80790     12.55     <.0001            0
    UTL         1     -0.42057      0.07116     -5.91     <.0001      3.53617
    SPA         1     -3.87295      0.83640     -4.63     <.0001      1.87192
    ALF         1     -6.40946      0.98175     -6.53     <.0001      1.43595
    TYPE        1     -3.53349      0.74134     -4.77     <.0001     46.03858
    utltp       1      0.42281      0.08682      4.87     <.0001     56.16250

          Test UTLTYPE1 Results for Dependent Variable CPM
                                   Mean
          Source          DF      Square    F Value    Pr > F

          Numerator        1    0.00018545     0.00    0.9653
          Denominator     27    0.09622
```

The deletion of the two product variables has virtually no effect on the fit of the model, and all coefficients are highly significant.

This equation estimates that for the short-haul airlines a one-unit (percent) change in utilization decreases cost by 0.42 cents. On the other hand, for the long-haul lines the corresponding effect is the sum of the coefficients for UTL and UTLTP; this estimate is $-0.420572 + 0.422809 = 0.002237$, or almost zero. A test that this effect is zero is performed by the TEST statement, and it appears that utilization is not a cost factor for these lines.

The use of dummy variables is readily extended to more than one categorical variable and also to situations where such variables have more than two categories. PROC GENMOD is a SAS procedure that will generate such dummy variables for variables with any number of categories. However, for all but the two-category cases (such as the one presented here) this type of analysis produces singular normal equations that must be solved by special methods. For this reason such models are more readily analyzed by PROC GLM, which automatically generates the dummy variables and provides for useful solutions of the normal equations. In addition, product variables can be specified directly with PROC GLM. The use of PROC GLM is presented in *SAS System for Linear Models, Third Edition.*

6.5 Binary Response Variable: Logistic Regression

In many regression applications the response variable has only two outcomes: an event either did or did not occur. Such a variable is often referred to as a *binary* or *binomial* variable as its behavior is related to the binomial distribution. A regression model with this type of response can be interpreted as a model that estimates the effect of the independent variables on the *probability* of the event occurring.

Binary response data typically appear in one of two ways:

❑ When observations represent individual subjects, the response is represented by a dummy or indicator variable having any two values. The most commonly used values are zero if the event does not occur and unity if it does.

❑ When observations summarize the occurrence of events for each set of unique combinations of the independent variables, the response variable is x/n where x is the number of occurrences and n the number of observations in the set.

Regression with a binary response is illustrated with data from a study of carriers of muscular dystrophy. Two groups of women, one consisting of known carriers of the disease and the other a control group, were examined for four types of protein in their blood. It is known that proteins may be used as a screening tool to identify carriers. The variables in the resulting data set are

CARRIER   coded 0 for control and 1 for carriers

P1   measurement of protein type 1, a type traditionally used for screening

P2   measurement of protein type 2

P3   measurement of protein type 3

P4   measurement of protein type 4

The objective is to determine the effectiveness of these proteins to identify carriers of the disease, with special reference on how screening is improved by using measurements of the other proteins. The data are shown in Output 6.14.

Output 6.14
Muscular
Dystrophy
Screening
Data

Obs	P1	P2	P3	P4	CARRIER
1	167	89.0	25.6	364	1
2	104	81.0	26.8	245	1
3	30	108.0	8.8	284	1
4	44	104.0	17.4	172	1
5	65	87.0	23.8	198	1
6	440	107.0	20.2	239	1
7	58	88.2	11.0	259	1
8	129	93.1	18.3	188	1
9	104	87.5	16.7	256	1
10	122	88.5	21.6	263	1
11	265	83.5	16.1	136	1
12	285	79.5	36.4	245	1
13	25	91.0	49.1	209	1
14	124	92.0	32.2	298	1
15	53	76.0	14.0	174	1
16	46	71.0	16.9	197	1
17	40	85.5	12.7	201	1
18	41	90.0	9.7	342	1
19	657	104.0	110.0	358	1
20	465	86.5	63.7	412	1
21	485	83.5	73.0	382	1
22	168	82.5	23.3	261	1
23	286	109.5	31.9	260	1
24	388	91.0	41.6	204	1
25	148	105.2	18.8	221	1
26	73	105.5	17.0	285	1
27	36	92.8	22.0	308	1
28	19	100.5	10.9	196	1
29	34	98.5	19.9	299	1
30	113	97.0	18.8	216	1
31	57	105.0	12.9	155	1
32	78	118.0	15.5	212	1
33	52	83.5	10.9	176	0
34	20	77.0	11.0	200	0
35	29	86.5	13.2	171	0
36	30	104.0	22.6	230	0
37	40	83.0	15.2	205	0
38	24	78.8	9.6	151	0
39	15	87.0	13.5	232	0
40	22	91.0	17.5	198	0
41	42	65.5	13.3	216	0
42	130	80.3	17.1	211	0
43	48	85.2	22.7	160	0
44	31	86.5	6.9	162	0
45	47	53.0	14.6	131	0
46	36	56.0	18.2	105	0
47	24	57.5	5.6	130	0
48	34	92.7	7.9	140	0
49	38	96.0	12.6	158	0
50	40	104.6	16.1	209	0
51	59	88.0	9.9	128	0
52	75	81.0	10.1	177	0
53	72	66.3	16.4	156	0
54	42	77.0	15.3	163	0
55	30	80.2	8.1	100	0
56	24	87.0	3.5	132	0
57	26	84.5	20.7	145	0
58	65	75.0	19.9	187	0
59	34	86.3	11.8	120	0
60	37	73.3	13.0	254	0
61	73	57.4	7.4	107	0
62	87	76.3	6.0	87	0
63	35	71.0	8.8	186	0
64	31	61.5	9.9	172	0
65	62	81.0	10.2	181	0
66	48	79.0	16.8	182	0
67	40	82.5	6.4	151	0
68	55	85.5	10.9	216	0
69	32	73.8	8.6	147	0
70	26	79.3	16.4	123	0

Because P1 has been the standard, it will be used to illustrate binomial regression with a single independent variable. Because this looks like a regression problem, you may first try a linear regression model with the following statements:

```
proc reg data=dystro;
   model carrier = p1;
run;
```

which provides the results shown in Output 6.15.

Output 6.15
Linear
Regression
with a Binary
Response

```
                            The REG Procedure
                              Model: MODEL1
                         Dependent Variable: CARRIER

                            Analysis of Variance

                                  Sum of        Mean
     Source            DF         Squares       Square     F Value    Pr > F

     Model              1         3.82804       3.82804      19.22    <.0001
     Error             68        13.54338       0.19917
     Corrected Total   69        17.37143


                 Root MSE              0.44628    R-Square     0.2204
                 Dependent Mean        0.45714    Adj R-Sq     0.2089
                 Coeff Var            97.62414


                            Parameter Estimates

                          Parameter     Standard
     Variable     DF       Estimate        Error     t Value    Pr > |t|

     Intercept     1        0.27462      0.06766        4.06      0.0001
     P1            1        0.00188      0.00042832      4.38     <.0001
```

The regression is certainly significant, and the estimated coefficients suggest that the probability of detecting a carrier increases with measurements of P1. You can obtain a plot of the fitted line curve and prediction intervals by adding the statement

```
plot carrier*p1/ pred;
run;
```

which provides the results shown in Output 6.16.[4]

---

[4] The lines for the predictions and confidence intervals have different colors on a computer monitor.

Output 6.16
Plot of
Linear
Response

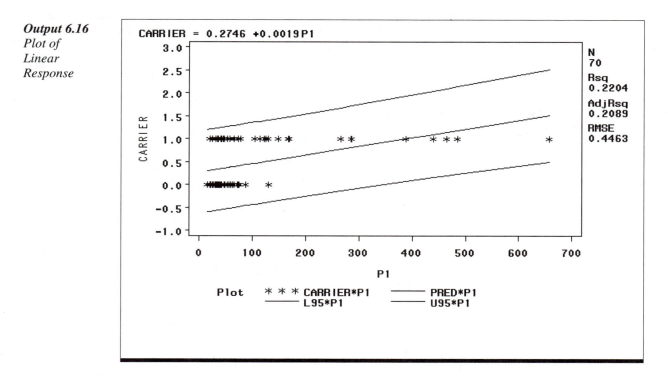

The plot immediately reveals a problem: the response variable is defined to be a set of probabilities that, by definition, are constrained to lie between zero and one, yet many estimated values and a large portion of the 95% prediction intervals are beyond this range.

Another difficulty with this model is that the variance of the binomial response variable is known to be a function of $p(1-p)$, where p is the probability of the event. This obviously violates the equal variance assumption required by the least squares estimation process. Thus this particular approach to the regression with a binary response appears to have limited usefulness.

The use of weighted regression may alleviate the unequal variance violation, and the use of the arcsine transformations may provide somewhat better estimates. However, a more useful approach is afforded by the *logistic* regression model. For a single independent variable the model is defined as

$$\log\left(\frac{p}{1-p}\right) = \beta_0 + \beta_1 x + \varepsilon,$$

where p is the probability of a response. The resulting response curve asymptotically approaches zero at one end and unity on the other (depending on the sign of β_1) and thus does not exhibit the difficulty found with the linear response function.

Note that although the regression model is linear on the right side, the left side is a nonlinear function of the response variable p. This function is known as the *logit link* function. And because it is not linear, the usual least squares methods cannot be used to estimate the parameters. Instead, a method known as *maximum likelihood* is used to obtain these estimates.

Given a model specified by a set of parameters and the distribution of the random error, it is possible to obtain a function that describes the likelihood of a sample arising from that model. This expression is used to find those values of the parameters that maximize the likelihood. These estimates are called maximum likelihood estimates. Also, since the model includes the specification of the error variance, the equal variance assumption is not required.

For linear regression models with normally distributed errors, the maximum likelihood principle produces a set of linear equations whose solution provides the familiar least squares estimates. However, for the logistic regression model, the equations that must be solved to obtain estimates of the parameters are not linear; hence they must be obtained by numerical methods. This means that there are no formulas for the partitioning of sums of squares and variances of estimates and the subsequent test statistics and confidence intervals. However, $-2*$logarithm of the likelihood provides a sample statistic that is related to the χ^2 distribution, which can be used for hypothesis testing. Note that because of the minus sign, smaller values signify higher likelihood, implying a better-fitting model.

The sampling distributions of the parameter estimates are obtained by the use of asymptotic theory, which means that, strictly speaking, they are valid only for infinite sample sizes. However, studies show that they usually do quite well for moderate sample sizes, although this is not guaranteed. One result of this is that slightly different approaches in the derivations of these variances will yield different test statistics. Software companies, such as SAS Institute, are obliged to offer these various statistics in their outputs, which often makes them confusing, especially for those who are not experts in the methodology.

In the SAS System, logistic regression can be performed by PROC LOGISTIC, which you can use much as you do PROC REG, although the results will look quite different.

To perform a logistic regression to estimate the probability of being a carrier as related to P1, use the following statements:

```
proc logistic data=dystro descending;
   model carrier = p1;
run;
```

The PROC option DESCENDING is needed because, by default, PROC LOGISTIC models the probability that CARRIER = 0, while the interest here is to model the probability that CARRIER = 1. The results are shown in Output 6.17. The circled numbers have been added to key the descriptions that follow.

Output 6.17
PROC
LOGISTIC
Output

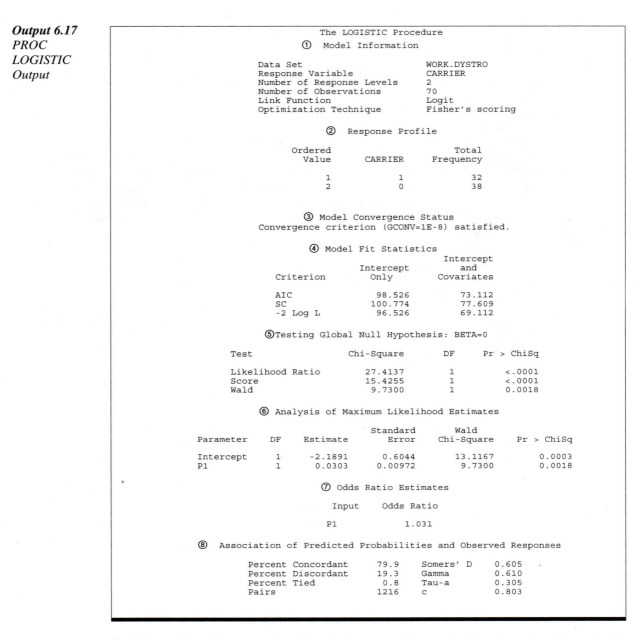

```
                        The LOGISTIC Procedure
                    ①  Model Information

            Data Set                        WORK.DYSTRO
            Response Variable               CARRIER
            Number of Response Levels       2
            Number of Observations          70
            Link Function                   Logit
            Optimization Technique          Fisher's scoring

                     ②  Response Profile

                 Ordered                       Total
                 Value        CARRIER        Frequency

                    1            1               32
                    2            0               38


                 ③ Model Convergence Status
         Convergence criterion (GCONV=1E-8) satisfied.

                 ④ Model Fit Statistics
                                              Intercept
                                Intercept        and
            Criterion             Only        Covariates

            AIC                   98.526        73.112
            SC                   100.774        77.609
            -2 Log L              96.526        69.112

          ⑤Testing Global Null Hypothesis: BETA=0

         Test              Chi-Square      DF      Pr > ChiSq

         Likelihood Ratio    27.4137        1        <.0001
         Score               15.4255        1        <.0001
         Wald                 9.7300        1        0.0018

          ⑥ Analysis of Maximum Likelihood Estimates

                                   Standard      Wald
         Parameter   DF   Estimate   Error    Chi-Square   Pr > ChiSq

         Intercept    1    -2.1891   0.6044    13.1167       0.0003
         P1           1     0.0303   0.00972    9.7300       0.0018

                   ⑦ Odds Ratio Estimates

                 Input      Odds Ratio

                  P1           1.031

    ⑧  Association of Predicted Probabilities and Observed Responses

            Percent Concordant    79.9    Somers' D    0.605
            Percent Discordant    19.3    Gamma        0.610
            Percent Tied           0.8    Tau-a        0.305
            Pairs                 1216    c            0.803
```

① This portion of the output describes the data set and some details on the estimation method. Link Function and Optimization Technique show the default link function and optimization method used by PROC LOGISTIC. Other options are available that may be useful for special applications.

② Response Profile describes the coded responses, which in this example are 0 for control and 1 for carrier. PROC LOGISTIC accepts both numeric and character values for this variable.[5] It is useful to inspect this table to verify that the response levels representing the events are correctly chosen. In this example CARRIER = 1 is the event whose probability is being modeled.

---

[5] In fact, PROC LOGISTIC can be used for ordinal response variables.

③ Model Convergence Status indicates that the numerical solution method did converge. If this message is different, the results of the analysis may be of questionable value.

④ Model Fit Statistics measures the fit of the model to the data. The magnitudes of the differences between the statistics for Intercept Only and Intercept and Covariates indicate how much better the model fits. The last line (-2 Log L) is used for the Likelihood Ratio test in ⑤ (below). The AIC and SC statistics adjust the likelihood statistics for sample size and the number of parameters and can be used to compare models.

⑤ The test of the global hypothesis is simply the test for the model, that is, the test that all coefficients are zero. In this case it is the test $\beta_1 = 0$. The Likelihood Ratio test is the difference in (-2 Log L) between the intercept-only and the full model. The resulting model Chi Square test statistic provides a p value of < 0.0001 for that test. The Score and Wald tests are alternative tests of the same hypothesis.

⑥ This portion provides the estimated parameters and associated tests. The parameters themselves are used for estimating probabilities using the formula

$$\hat{p} = 1/(1 + e^{-\hat{y}}),$$

where $\hat{y}$ is the responses obtained from the estimated parameters. They are also used to obtain the odds ratio estimate (see item ⑦). The estimates are available in the OUTPUT data set. If there are several independent variables, the individual tests are, of course, useful for determining the importance of the individual variables.

⑦ The odds ratio, computed as e^{β_1} is the change in the event odds, defined as

$$\frac{\text{P(event)}}{\text{P(nonevent)}}$$

for a unit increase in the independent variable. In this example the estimated odds ratio is $e^{0.0303} = 1.031$; hence the odds of being classified as a carrier increase by 3.1% for each unit increase in P1.

⑧ This portion provides measures of association to assess the quality of the model. These measures are based on an analysis of individual pairs of observations with different responses.

In this example there are 38 zeroes and 32 ones; hence there are $38 \times 32 = 1216$ such pairs. A pair is deemed *concordant* if the observation with the higher response also has the higher estimated probability, *discordant* if the reverse is true, and *tied* if the estimated probabilities are identical. The numbers given are the percentage of pairs in each of these classes; obviously, the higher the percentage of concordant pairs, the better is the fit of the model.

The right-hand side of this portion of the output gives four different measures of rank correlation computed from these quantities. These correlations may range from zero to one; hence a larger correlation implies a stronger relationship. These correlations may be used to compare different models.

Unlike PROC REG, PROC LOGISTIC does not have interactive plotting capabilities, but the procedure does have the capability of producing a data set that contains variables suitable for plotting. This is produced by an OUTPUT statement, which has the same general format as that for PROC REG.

To produce such a data set and plot the estimated probabilities and confidence intervals, use the following statements:

```
proc logistic data=dystro descending;
        model carrier = p1;
output out=a p=pcarrier l=lower u=upper;
proc sort; by p1;
proc gplot data=a;
        symbol1 v=star c=black;
        symbol2 v=point i=spline l= 1 c=black;
        symbol3 v=point i=spline l= 33 c=black ;
        plot carrier*p1=1 pcarrier*p1= 2 (upper lower)*p1=3 /
        overlay;
run;
```

The OUTPUT statement requests the predicted values and the 95% confidence limits on the probabilities; other confidence levels may be requested with the ALPHA option. PROC SORT is required to obtain the smooth lines joining the individual points. The resulting plot is shown in Output 6.18.

Output 6.18
Estimates
from PROC
LOGISTIC

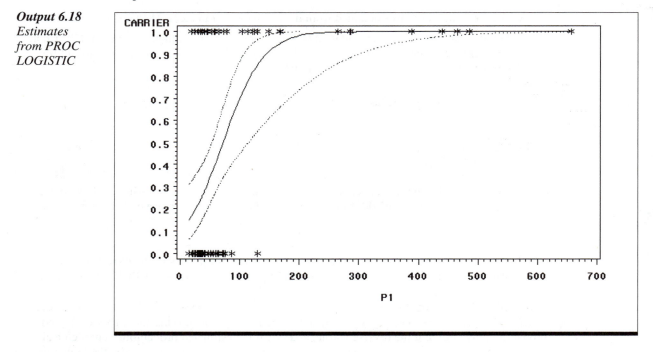

Note that the estimates and confidence bounds stay within the [0,1] range, as probabilities should.

As noted, data for a binomial response variable may be grouped. That is, an observation for a specific value of the independent variable will indicate that *x* out of *n* observations are a success. PROC LOGISTIC can analyze data of this form as follows:

```
proc logistic;
   model x/n = p1 /descending;
run;
```

Other options and statements are not affected by using this form of the model.

In addition to being able to perform a logistic regression model with several variables, PROC LOGISTIC can also perform variable selection procedures that are implemented in a manner essentially identical to PROC REG. These include SELECTION=FORWARD, BACKWARD, and STEPWISE. The option SELECTION=SCORE is essentially the same as SELECTION=RSQUARE in PROC REG. All use the same syntax used by PROC REG. The step-type procedures provide abbreviated outputs for each step, and the Score method produces a listing of included variables for each model and the associated Score fit statistic.

To perform the Score selection use the following statements:

```
proc logistic data=dystro;
   model carrier = p1 p2 p3 p4 / selection=score
   best=4;
run;
```

The results are shown in Output 6.19.

Output 6.19
Variable
Selection
with PROC
LOGISTIC

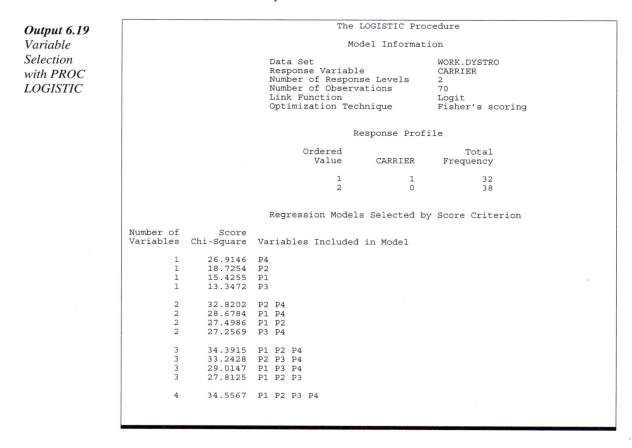

```
                        The LOGISTIC Procedure

                          Model Information

              Data Set                     WORK.DYSTRO
              Response Variable            CARRIER
              Number of Response Levels    2
              Number of Observations       70
              Link Function                Logit
              Optimization Technique       Fisher's scoring


                          Response Profile

                   Ordered                        Total
                    Value      CARRIER          Frequency

                       1          1                  32
                       2          0                  38


            Regression Models Selected by Score Criterion

   Number of       Score
   Variables    Chi-Square   Variables Included in Model

          1       26.9146    P4
          1       18.7254    P2
          1       15.4255    P1
          1       13.3472    P3

          2       32.8202    P2 P4
          2       28.6784    P1 P4
          2       27.4986    P1 P2
          2       27.2569    P3 P4

          3       34.3915    P1 P2 P4
          3       33.2428    P2 P3 P4
          3       29.0147    P1 P3 P4
          3       27.8125    P1 P2 P3

          4       34.5567    P1 P2 P3 P4
```

The choice of the most appropriate model is somewhat subjective, much as it was for the results of the R-Square selection method. The step-type selection methods do provide *p* values, but these are, of course, of limited use. Because larger values of the Score statistic imply better fit, changes in that statistic may be used as a guide. The decrease in the Score statistic from deleting P3 is quite small, suggesting that P3 should be eliminated. Deleting P1 (the standard) generates a somewhat greater decrease, and the deletion of P2 results in a major decrease. The two-variable model has the advantage that the best model is quite superior to the second-best model.

You can now examine the results for the best two-variable model with the statements

```
proc logistic data=dystro;
   model carrier = p2 p4;
run;
```

The results are shown in Output 6.20.

Output 6.20
PROC
LOGISTIC for
P2 and P4

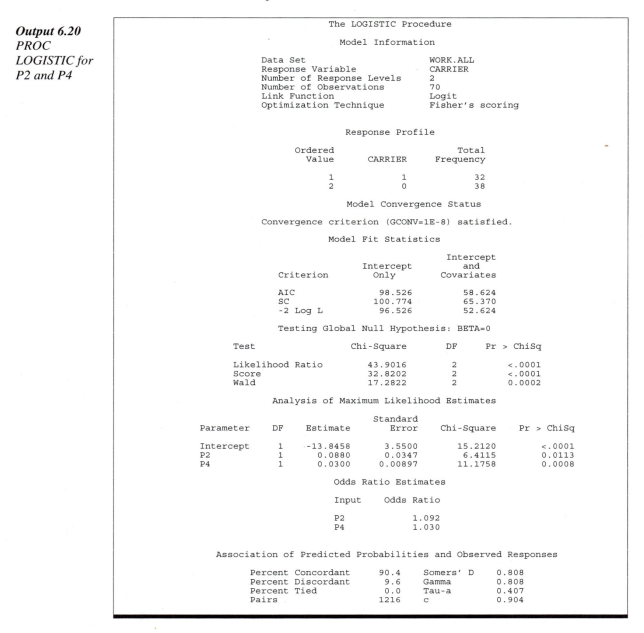

```
                         The LOGISTIC Procedure

                            Model Information

          Data Set                        WORK.ALL
          Response Variable               CARRIER
          Number of Response Levels       2
          Number of Observations          70
          Link Function                   Logit
          Optimization Technique          Fisher's scoring


                            Response Profile

              Ordered                            Total
               Value        CARRIER            Frequency

                 1             1                   32
                 2             0                   38

                       Model Convergence Status

          Convergence criterion (GCONV=1E-8) satisfied.

                         Model Fit Statistics

                                                Intercept
                                   Intercept       and
                 Criterion           Only       Covariates

                 AIC                 98.526        58.624
                 SC                 100.774        65.370
                 -2 Log L            96.526        52.624

                 Testing Global Null Hypothesis: BETA=0

          Test                Chi-Square      DF      Pr > ChiSq

          Likelihood Ratio      43.9016        2        <.0001
          Score                 32.8202        2        <.0001
          Wald                  17.2822        2        0.0002

               Analysis of Maximum Likelihood Estimates

                                  Standard
          Parameter   DF   Estimate    Error   Chi-Square   Pr > ChiSq

          Intercept    1   -13.8458    3.5500    15.2120      <.0001
          P2           1     0.0880    0.0347     6.4115      0.0113
          P4           1     0.0300    0.00897   11.1758      0.0008

                           Odds Ratio Estimates

                          Input      Odds Ratio

                           P2          1.092
                           P4          1.030


          Association of Predicted Probabilities and Observed Responses

                Percent Concordant      90.4     Somers' D    0.808
                Percent Discordant       9.6     Gamma        0.808
                Percent Tied             0.0     Tau-a        0.407
                Pairs                   1216     c            0.904
```

The two-variable model does provide a better fit than that obtained from the standard protein P1, especially as indicated by the increase in the rank order statistic. Both coefficients are positive and statistically significant.

The estimated response curve for the two-variable model can be plotted in a manner similar to that used for response surfaces (Chapter 5). Use the following statements:

```
data grid;
      do p2 = 50 to 100 by 5;
      do p4 = 140 to 340 by 20;
      output; end; end;
data all; set dystro grid;
proc logistic data=all descending;
      model carrier = p2 p4;
output out=a p=pcarrier;
data plot; set a;
      if carrier=.;
proc g3d data=plot;
      plot p2*p4=pcarrier/ rotate=30;
run;
```

The resulting plot is shown in Output 6.21.

Output 6.21
Two Variable
Logistic
Response
Curve

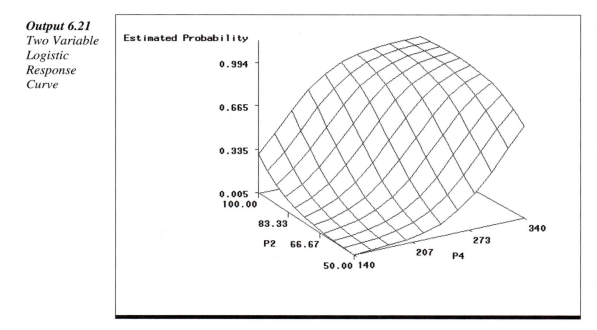

PROC LOGISTIC has a large number of additional features, for example:

❑ modeling an ordinal response variable

❑ using different link functions (e.g., probit model)

❑ fine-tuning the iterative solution process

❑ computing different fit statistics

❑ creating an OUTEST data set containing parameter estimates

❑ generating influence statistics that may be output to a data set and plotted.

6.6 Summary

The purpose of this chapter has been to show how linear regression models can be adapted to many situations. PROC REG can be used to fit a wide variety of models that do not look linear. Using the logarithmic transformation on the dependent variable stabilizes the error variance, and then using the logarithms of the independent variables provides a model whose coefficients are very useful in many applications. The use of specially coded variables allows the fitting of segmented regression or spline models that describe relationships where changes occur too abruptly to be fitted by polynomial models. Finally, indicator variables provide models that describe different relationships for different portions of the data.

Another class of models regresses a linear function of the independent variables on functions of the response variable. An important application of this type of model is the logistic model, which is appropriate for modeling binomial and ordinal responses. Such models are fitted by PROC LOGISTIC. A number of other response variables may be fitted with PROC GENMOD, which can also perform analysis of variance and covariance models in a manner similar to that of PROC GLM.

Nonlinear Models

7.1 Introduction

The term *linear model* refers to a model that is linear in the parameters or that can be made linear by transformation (such as the use of logarithms to estimate a multiplicative model) or redefinition of variables (see the BOQ data in Chapter 4, "Multicollinearity: Detection and Remedial Measures"). It is apparent that such linear models need not be linear in the variables. However, many relationships exist that cannot be described by linear models or adaptations of linear models. For example, the exponential decay model

$$y = \beta e^{\gamma t} + \varepsilon$$

is not linear in its parameters. Specifically, the term $\beta e^{\gamma t}$ is not a linear function of γ. This particular nonlinear model, called the exponential growth or decay model, is used to represent increase (growth) or decrease (decay) over time (t) of many types of responses such as population size or radiation counts.

One major advantage of many nonlinear models over, say, polynomial models, is that the parameters represent meaningful physical quantities of the process described by the model. In the above model, the parameter β is the initial value of the response (when $t = 0$) and the parameter γ is the rate of exponential growth (or decay). A positive value of γ indicates growth while a negative value indicates decay. Additionally, if it is a decay model, the parameters may be used to estimate the half-life, that is, the time it takes for half of the substance to decay.

When a model is nonlinear in the parameters, the entire process of estimation and statistical inference is radically altered. This happens mainly because the normal equations that are solved to obtain least-squares parameter estimates are themselves nonlinear. Solutions of systems of nonlinear equations are not usually available in closed form (mathematical jargon for not being in the form of an equation), but they must be obtained by numerical methods. For this reason, closed-form expressions for the partitioning of sums of squares, as well as the consequently obtained statistics for making inferences on the parameters, are also unavailable.

For most applications, the solutions to the normal equations are obtained by means of an iterative process. The process starts with some preliminary estimates of the parameters. These estimates are used to calculate a residual sum of squares, and additional calculations are performed to give an indication of modifications of the parameter estimates that may reduce the residual sum of squares. This process is repeated until it appears that no further modification of parameter estimates results in a reduction of the residual sum of squares.

The SAS procedure for analyzing nonlinear models is the NLIN procedure. PROC NLIN is introduced with an example of the simple exponential decay model in Section 7.2, "Estimating the Exponential Decay Model." Additional examples of the use of PROC NLIN include the logistic growth model (see Section 7.3, "Fitting a Growth Curve with the NLIN Procedure") and the estimation of spline functions when the knot is estimated from the data (see Section 7.4, "Fitting Splines with Unknown Knots").

These three models are relatively simple and the estimation of their parameters is relatively straightforward. This is certainly not true of all nonlinear models, and for this reason PROC NLIN features a number of options to facilitate estimation of more complex models. Section 7.5, "Additional Comments on the NLIN Procedure" provides an overview of a number of these special options. [1]

7.2 Estimating the Exponential Decay Model

The SAS data set DECAY comes from an experiment to determine the radioactive decay of a substance. The variable COUNT represents the radiation count recorded at various times (TIME). Output 7.1 shows the data, and Output 7.2 gives the plot of COUNT versus TIME. The decrease (decay) of COUNT with TIME is clearly evident.

Output 7.1
Radiation
Count Data

Obs	TIME	COUNT
1	0	383.0
2	14	373.0
3	43	348.0
4	61	328.0
5	69	324.0
6	74	317.0
7	86	307.0
8	90	302.0
9	92	298.0
10	117	280.0
11	133	268.0
12	138	261.0
13	165	244.0
14	224	200.0
15	236	197.0
16	253	185.0
17	265	180.0
18	404	120.0
19	434	112.5

---

[1] Nonlinear models may also be fitted using the MODEL procedure in SAS/ETS software.

Output 7.2
Plot of
Radiation
Counts

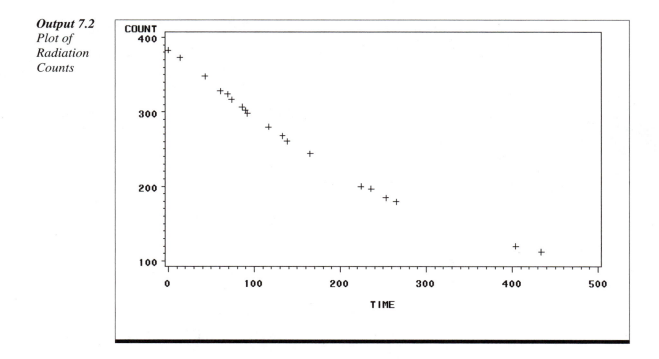

You use PROC NLIN for estimating the exponential decay model

$$y = \beta e^{\gamma t} + \varepsilon$$

with the following statements:

```
proc nlin data=decay;
   parms b=380 c=-0.0026;
   model count = b*exp(c*time);
run;
```

where B and C are used to represent the parameters β and γ, respectively.

You can see the differences between executing PROC NLIN and PROC REG.

❏ In PROC NLIN you must specify the complete model (except for the error) in the MODEL statement, whereas in PROC REG you need to list only the names of dependent and independent variables.

❏ In PROC NLIN you must identify the names of the parameters in the PARMS statement because the MODEL statement includes both variables and parameters.

❏ The PARMS statement is also used to provide starting values of the parameters that initiate the iterative estimation procedure. In this case, the starting values are specified as B = 380 and C = − 0.0026.

Providing good starting values is quite important because poor starting values can increase computing time and may even prevent finding correct estimates of the parameters. Starting values are usually educated guesses, although some preliminary calculations may be used for this purpose.

In this example, B is the expected COUNT at TIME = 0, which should be close to the observed value of 383. The initial value for C is obtained by using the observed count for a specific time and solving for C. Choose a value for TIME, say, 117, where COUNT = 280, and solve

$$280 = 380*e^{117*C}.$$

Taking logarithms makes this a linear equation, which is easily solved to obtain the initial value C = − 0.0026.

Finding initial values is not always straightforward. For such cases, PROC NLIN provides a grid search that may be useful (see Sections 7.3 and 7.5). The results from PROC NLIN appear in Output 7.3.

Output 7.3
Output from
PROC NLIN
for Radiation

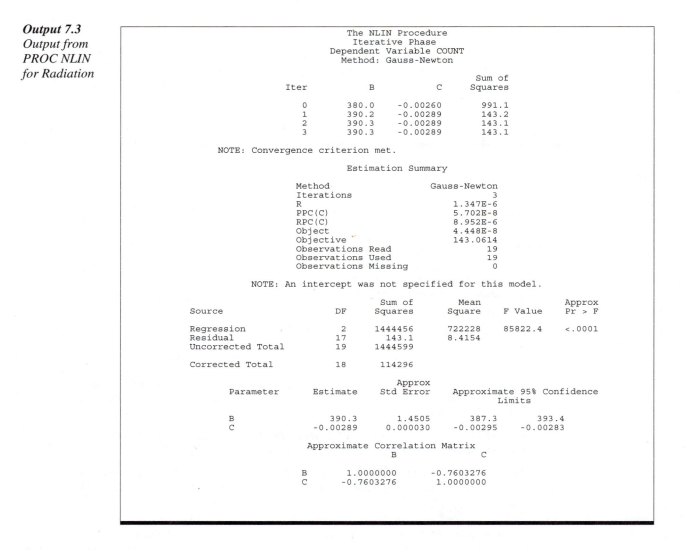

```
                        The NLIN Procedure
                         Iterative Phase
                    Dependent Variable COUNT
                      Method: Gauss-Newton

                                              Sum of
             Iter          B          C       Squares

              0         380.0     -0.00260     991.1
              1         390.2     -0.00289     143.2
              2         390.3     -0.00289     143.1
              3         390.3     -0.00289     143.1

     NOTE: Convergence criterion met.

                        Estimation Summary

             Method                    Gauss-Newton
             Iterations                           3
             R                           1.347E-6
             PPC(C)                      5.702E-8
             RPC(C)                      8.952E-6
             Object                      4.448E-8
             Objective                   143.0614
             Observations Read                 19
             Observations Used                 19
             Observations Missing               0

       NOTE: An intercept was not specified for this model.

                         Sum of         Mean                Approx
    Source          DF   Squares       Square     F Value   Pr > F

    Regression       2   1444456       722228     85822.4   <.0001
    Residual        17     143.1       8.4154
    Uncorrected Total 19  1444599

    Corrected Total 18    114296

                              Approx
    Parameter     Estimate    Std Error   Approximate 95% Confidence
                                                    Limits

        B          390.3       1.4505       387.3        393.4
        C        -0.00289      0.000030    -0.00295     -0.00283

                   Approximate Correlation Matrix
                         B                   C

            B      1.0000000          -0.7603276
            C     -0.7603276           1.0000000
```

The top portion of the output summarizes the iterative solution process. The method used here is the default Gauss-Newton method, and the following lines show that only four iterations were needed for the iterative search method to converge. In fact, the first step was almost sufficient. This is partly due to the rather good starting values of a relatively simple model.

The next section provides some technical details of the iterative process, which are primarily of interest if there was a problem with the iterative process.

In the next portion of the output, you find the partitioning of the sum of squares that corresponds to the ANOVA portion of the PROC REG output. The regression and residual sums of squares and mean squares have the same interpretations as in PROC REG, although in nonlinear models the regression sum of squares is computed directly from the predicted values because there is no shortcut formula.

However, note that the partitioning starts with the *uncorrected* total sum of squares. The reason for this is that in most nonlinear models there is no natural mean or intercept; hence the corrected sum of squares may have no meaning. The *F* value shown is then the test for the regression starting with the uncorrected total sum of squares. Note, however, that the corrected total sum of squares is also provided in case you want to compute the equivalents of the usual *F* test or R-Square statistic. You should be aware that the resulting R-Square might not have the usual interpretation (Kvålseth 1985).

In the final portion of the output, there is information on the estimated coefficients. First, you see the parameter estimates, which provide the estimated model

$$\widehat{\text{COUNT}} = 390.34 * e^{-0.00289 * \text{TIME}}.$$

The estimated initial count is 390.34 and the estimated exponential decay rate is -0.00289. This means that the expected count at time t is $e^{0.00289} = 0.997$ times the count at time $t-1$. In other words, the estimated rate of decay is $(1 - 0.997) = 0.003$, or approximately 0.3% per time period.

The estimated decay rate coefficient is used to get an estimated half-life, the time at which one half of the radiation has occurred. This is computed as $T2 = \ln(2)/-0.00289 = 240$ time periods.

The standard errors of the estimated coefficients and the confidence intervals are *asymptotic*. This means that the formulas used for the computations are only approximately correct because they are based on mathematical theory that is valid only for very large sample sizes. So you can be approximately 95% confident that the true exponential decay rate is between -0.00295 and -0.00283.

The final portion shows the approximate correlations between the estimated coefficients. You can see that there exists a moderately large negative correlation between the estimates of the two coefficients.

Unlike PROC REG, PROC NLIN does not have interactive plotting capabilities, so you must create a data set with predicted and residual values for plotting and otherwise checking assumptions. Creating a data set is done as it is with PROC REG. Use the following statements:

```
proc nlin data=decay;
   parms b=380 c=-0.0026;
   model count = b*exp(c*time);
output out=plot p=pct r=rct;
proc plot data=plot hpercent=50 vpercent=50;
   plot rct*time / vref=0;
run;
```

The resulting plot is shown in Output 7.4. The only outstanding feature is the rather large negative residual for the first observation. A box plot for these residuals (not shown) does indicate that this is a mild outlier.

Output 7.4
Residual
Plot

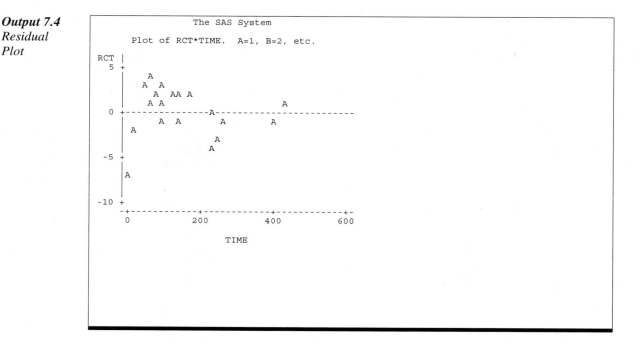

You can use the output data set to plot the observed values and the predicted curve. The statements are

```
symbol1 v=star c=black;
symbol2 v=point c=black i=spline l=1;
proc gplot data =plot;
plot count*time=1 pct*time=2 / overlay;
run;
```

The plot is shown in Output 7.5.

Output 7.5
Plot of
Predicted
Curve

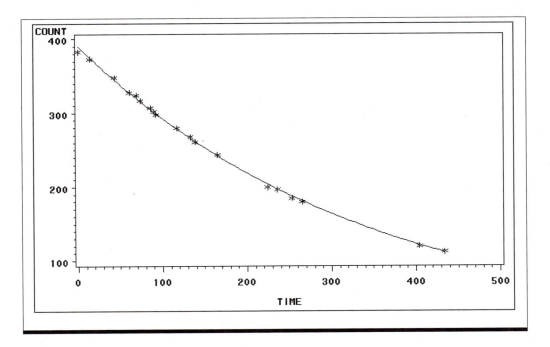

The curve does seem to fit rather well but you can also see that the relatively large negative residual for the first observation appears to be due to an almost linear decay for the first three time periods.

Although most nonlinear models must be fitted by an iterative procedure such as PROC NLIN, some may be linearized and the linearized versions fitted by linear regression. The deterministic portion of the above decay model is multiplicative and can be linearized by taking logarithms on both sides of the equation and performing a linear regression using these values. In this example, the model

$$\log(\text{COUNT}) = \beta + \gamma(\text{TIME}) + \varepsilon$$

fits the decay model, with e^{β} being the estimated initial value and γ the decay constant. You can do this using the following statements:

```
data logdecay; set decay;
   logcount = log(count);
proc reg data = logdecay;
   model logcount = time;
run;
```

Output 7.6
Fitting a
Linearized
Nonlinear
Model

```
                              The REG Procedure
                               Model: MODEL1
                          Dependent Variable: logcount

                             Analysis of Variance

                                    Sum of         Mean
        Source            DF       Squares        Square     F Value    Pr > F

        Model              1       2.25695       2.25695     20390.9    <.0001
        Error             17       0.00188    0.00011068
        Corrected Total   18       2.25883


                  Root MSE              0.01052    R-Square     0.9992
                  Dependent Mean        5.52535    Adj R-Sq     0.9991
                  Coeff Var             0.19041


                             Parameter Estimates

                           Parameter      Standard
        Variable    DF      Estimate         Error     t Value    Pr > |t|

        Intercept    1       5.96858       0.00393     1517.98     <.0001
        TIME         1      -0.00291    0.00002035     -142.80     <.0001
```

The result in Output 7.6 provides the estimate of initial value of $\exp(5.9686) = 390.95$ and decay constant of -0.00291, which compare favorably with the values 390.3 and -0.00289 obtained by PROC NLIN in Output 7.3. The standard error of the linearized model estimate of the decay constant is 0.00002035, compared with the asymptotic standard error of 0.0000297 from PROC NLIN. The differences arise primarily due to the fact that linearized estimates are least squares estimates for the logarithmic model and are intended for use when the standard deviation of residuals is proportional to the mean (see Section 6.2, "Multiplicative Models"). Since the residuals do not appear to have this feature (see Output 7.4), the linearized model may not be appropriate.

7.3 Fitting a Growth Curve with the NLIN Procedure

A common application of nonlinear regression is fitting growth curves. This application is illustrated with the fish growth data from Chapter 5, "Curve Fitting," using the data for fish kept at 29 degrees Celsius. For convenience, the data are shown here in Output 7.7, with the fish length denoted as the variable LEN29.

Output 7.7
Fish Growth
Data for
Temperature
of 29
Degrees

```
                    Obs    AGE    LEN29

                     1      14      590
                     2      21      910
                     3      28     1305
                     4      35     1730
                     5      42     2140
                     6      49     2725
                     7      56     2890
                     8      63     3685
                     9      70     3920
                    10      77     4325
                    11      84     4410
                    12      91     4485
                    13      98     4515
                    14     105     4480
                    15     112     4520
                    16     119     4545
                    17     126     4525
                    18     133     4560
                    19     140     4565
                    20     147     4626
                    21     154     4566
```

The relationship used here is known as the logistic growth curve.[2] The general form of the equation for the logistic growth curve is

$$y = \frac{k}{1 + ((k - n_0)/n_0)e^{-rt}} + \varepsilon \quad .$$

This model has three parameters: k, n_0, and r. The parameter n_0 is the expected value of y at time $t = 0$, k is the height of the horizontal asymptote (the expected value of y as t approaches infinity), and r is a measure of growth rate. The term ε is the random error and is assumed to have mean zero and variance σ^2.

You can fit the logistic model to the fish growth data with the following statements:

```
proc nlin data=fish;
    parms k=4500 no=500 r=1;
    model len29 = k/(1+((k-no)/no)*exp(-r*age));
run;
```

Note that AGE represents the variable t in the growth curve equation, and because you cannot have subscripts in SAS code, n_0 is represented by NO. The starting values for the parameters in the PARMS statement are preliminary guesses based on knowledge of what the parameters stand for. The value K = 4500 is selected because the values of LEN29 appear to level out at approximately 4500, and the value NO = 500 is selected because the value of LEN29 for early ages is around 500. Less is known about the value for R. Since growth is positive (the values of LEN29 increase with age), R should be a small positive number. Because the starting values for K and NO appear to be quite good, an arbitrarily chosen starting value of 1 should suffice for R. The results are shown in Output 7.8, except that the iteration summary has been omitted to save space.

---

[2] This is not to be confused with the logistic model for binary response. See Section 6.5, "Binary Response Variable: Logistic Regression."

Output 7.8
PROC NLIN
Output for
Logistic
Growth
Curve
Regression

```
                         Estimation Summary
                Method                   Gauss-Newton
                Iterations                       53
                Subiterations                   113
                Average Subiterations      2.132075
                R                           1.03E-6
                PPC(K)                     3.953E-7
                RPC(K)                      0.00012
                Object                     9.735E-8
                Objective                  38462029
                Observations Read                21
                Observations Used                21
                Observations Missing              0

                              Sum of        Mean                 Approx
     Source            DF     Squares       Square    F Value    Pr > F

     Regression         1     2.6088E8     2.6088E8
     Residual          20     38462029      1923101
     Uncorrected Total 21     2.9934E8

     Corrected Total   20     38462029

                          The NLIN Procedure

                              Approx
         Parameter    Estimate   Std Error   Approximate 95% Confidence
                                                        Limits

         K              3524.6       302.6    2893.4        4155.9
         N0          -4.84E32       Infty     -Infty         Infty
         R            1.379E28       Infty     -Infty         Infty

                   Approximate Correlation Matrix
                         K                N0                  R

            K       1.0000000             .                  .
            N0          .                 .                  .
            R           .                 .                  .
```

Because no options have been specified in the PROC NLIN statement, the default Gauss-Newton method has been used. The summary shows that there were 53 iterations, which is a rather large number for such a relatively simple function. Further examination of the output shows that the residual sum of squares is the same as the corrected total sum of squares, indicating that the fitted equation is the mean! Finally, the estimated coefficients for NO and R have infinite standard errors. Obviously the procedure has not worked, and the probable reason for this is bad starting values.

The starting values for K and NO were based on the observations and are probably quite good. However, the starting value of 1.0 for R was quite arbitrary, and it should probably be changed. At this point you can use the search capabilities of the PARMS statement. Using the statement

```
parms k=4500 no=500 r = .1 to 1 by .1;
```

causes PROC NLIN to evaluate the residual sum of squares for all of these values and then start the iterations for that set of values having the minimum residual sum of squares. The results are shown in Output 7.9.

Output 7.9
PROC NLIN
with Grid
Search

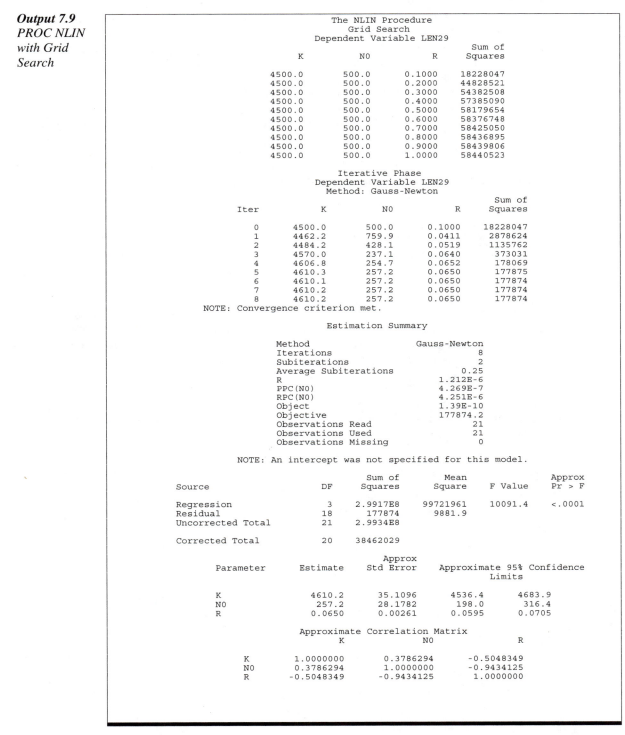

```
                          The NLIN Procedure
                            Grid Search
                        Dependent Variable LEN29
                                                Sum of
            K              N0            R      Squares

          4500.0         500.0        0.1000    18228047
          4500.0         500.0        0.2000    44828521
          4500.0         500.0        0.3000    54382508
          4500.0         500.0        0.4000    57385090
          4500.0         500.0        0.5000    58179654
          4500.0         500.0        0.6000    58376748
          4500.0         500.0        0.7000    58425050
          4500.0         500.0        0.8000    58436895
          4500.0         500.0        0.9000    58439806
          4500.0         500.0        1.0000    58440523

                          Iterative Phase
                       Dependent Variable LEN29
                        Method: Gauss-Newton
                                                Sum of
     Iter        K            N0          R     Squares

        0      4500.0       500.0      0.1000    18228047
        1      4462.2       759.9      0.0411     2878624
        2      4484.2       428.1      0.0519     1135762
        3      4570.0       237.1      0.0640      373031
        4      4606.8       254.7      0.0652      178069
        5      4610.3       257.2      0.0650      177875
        6      4610.1       257.2      0.0650      177874
        7      4610.2       257.2      0.0650      177874
        8      4610.2       257.2      0.0650      177874
NOTE: Convergence criterion met.

                        Estimation Summary

             Method                     Gauss-Newton
             Iterations                       8
             Subiterations                    2
             Average Subiterations          0.25
             R                            1.212E-6
             PPC(N0)                      4.269E-7
             RPC(N0)                      4.251E-6
             Object                       1.39E-10
             Objective                    177874.2
             Observations Read                21
             Observations Used                21
             Observations Missing              0

         NOTE: An intercept was not specified for this model.

                               Sum of        Mean                  Approx
     Source            DF      Squares       Square     F Value    Pr > F

     Regression         3     2.9917E8      99721961    10091.4    <.0001
     Residual          18       177874       9881.9
     Uncorrected Total 21     2.9934E8

     Corrected Total   20     38462029

                                  Approx
     Parameter     Estimate     Std Error    Approximate 95% Confidence
                                                        Limits

     K              4610.2       35.1096      4536.4        4683.9
     N0              257.2       28.1782       198.0         316.4
     R             0.0650        0.00261      0.0595        0.0705

                   Approximate Correlation Matrix
                        K             N0              R

     K           1.0000000      0.3786294      -0.5048349
     N0          0.3786294      1.0000000      -0.9434125
     R          -0.5048349     -0.9434125       1.0000000
```

The first portion of the output shows the results of the initial search procedure. It appears that the initial choice of R was indeed poor, and using 0.1 reduces the residual sum of squares by a factor of 3. The iterative procedure now uses only eight steps and decreases the residual mean square by a factor of 100. However, even with this reduction you should check to see if this is a reasonable value for the minimum residual sum of squares. To do this, compare the residual SS of 177874 from the PROC NLIN output with the residual SS of 211239 from the fourth-degree polynomial model fitted in Output 5.2. You see that the residual SS for the logistic fit is actually somewhat smaller than that for the fourth-degree polynomial fit. This indicates that the logistic model fits quite well, especially considering that the logistic model has only three parameters compared with five for the fourth-degree polynomial.

The rounded final parameter estimates give the fitted model

$$\widehat{LEN29} = \frac{4610}{1 + \left(\left(4610 - 257\right)/257\right)e^{0.065*AGE}}$$

$$= \frac{1}{0.00022 + 0.0037\,e^{-0.065*AGE}} .$$

Asymptotic standard errors and corresponding confidence intervals for the parameters show that they are estimated with useful precision. The large negative correlation between NO and R indicates the presence of the nonlinear version of multicollinearity. This could indicate a family of curves with differing values of R and NO that all fit about as well as each other, meaning that parameter estimates may be imprecise. However, the satisfactorily small standard errors indicate this is not the case.

You can examine the residuals from the fitted logistic growth curve as a further check on the fit of the model. Obtain a data set containing the residuals by adding the following statements:

```
output out=plotdata p=plen29 r=rlen29;
run;
```

Plot the residuals with the following statements:

```
proc gplot data=plotdata;
   plot rlen29*age/ vref=0;
run;
```

Output 7.10
Residual
Plot for
Logistic
Curve

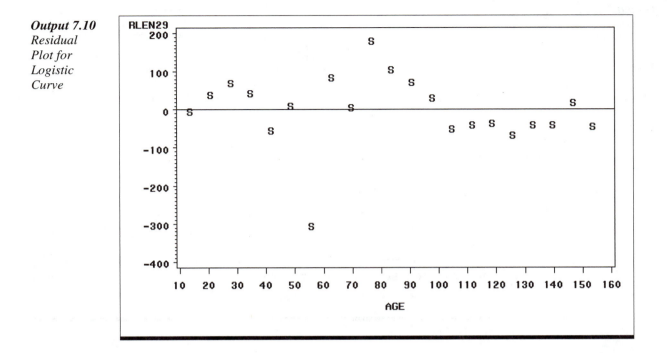

There is a large negative residual of about − 300 corresponding to an age in the upper 60s, which may be an outlier. Also, there is a string of six negative residuals from ages 105 to 140, indicating a modest lack of fit in this region.

Now plot the fitted curve through the data with the following statements:

```
proc gplot data=plotdata;
   symbol1 v=star c=black;
   symbol2 v=point c=black i=spline l=1;
   plot len29*age=1 plen29*age=2 / overlay;
run;
```

Results in Output 7.11 do indeed show the possible outlier and the poor fit at several adjacent points.

Output 7.11
Plot of
Actual and
Predicted
Values

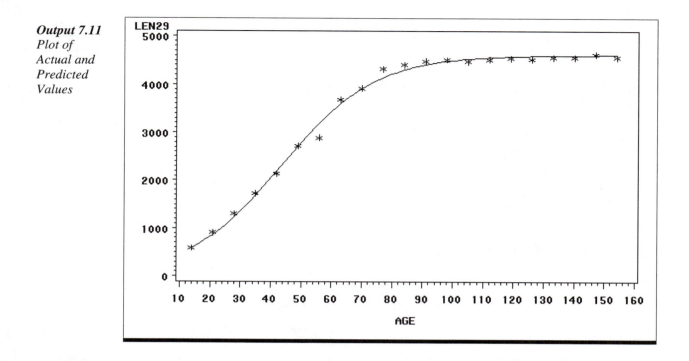

7.4 Fitting Splines with Unknown Knots

Section 6.3, "Spline Models," shows how PROC REG is used to fit a spline model when the knot is assumed known. Consider the more difficult problem of fitting the linear spline model with an estimated knot. This cannot be done with linear regression methods, because now the knot is a parameter of the model.

As noted earlier, the spline model can be represented as

$$\widehat{\text{LEN29}} = \beta_0 + \beta_1 * \text{AGE} + \beta_2 * \text{AGEPL} + \varepsilon$$

where AGEPL = AGE − KNOT if AGE ≥ KNOT, and AGEPL = 0 if AGE < KNOT. The variable AGEPL is equivalent to the variable AGEPLUS that was used when KNOT was fixed at 80 (Section 6.3). You can use PROC NLIN to fit the model with estimated knot, as follows:

```
proc nlin data=fish;
    parms b0=-328 b1=61 b2=-59 knot=80;
    agepl=max(age-knot, 0);
    model len29 = b0 +b1*age + b2*agepl;
run;
```

Starting values for the PARMS statement are those from the linear regression model fitted with fixed KNOT = 80 (Output 6.7).

Output 7.12
Estimating Linear Spline with PROC NLIN

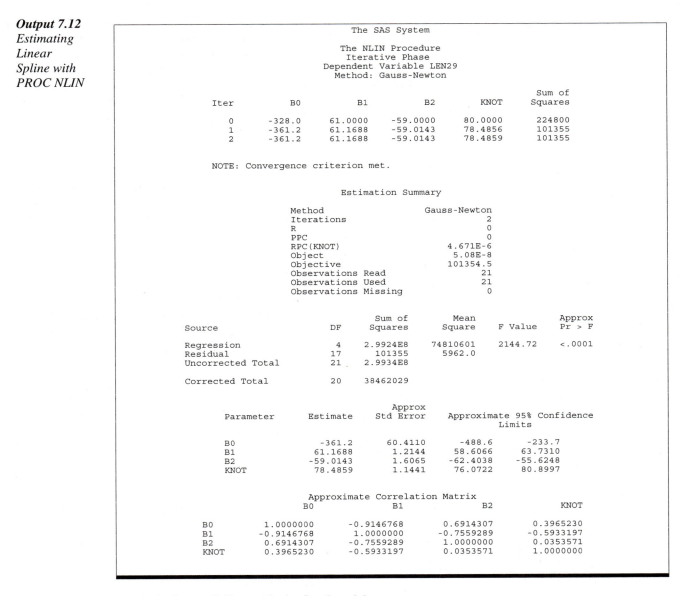

```
                              The SAS System

                            The NLIN Procedure
                             Iterative Phase
                        Dependent Variable LEN29
                          Method: Gauss-Newton

                                                          Sum of
    Iter        B0          B1          B2        KNOT    Squares

      0      -328.0     61.0000     -59.0000    80.0000    224800
      1      -361.2     61.1688     -59.0143    78.4856    101355
      2      -361.2     61.1688     -59.0143    78.4859    101355


    NOTE: Convergence criterion met.


                          Estimation Summary

                  Method                     Gauss-Newton
                  Iterations                            2
                  R                                     0
                  PPC                                   0
                  RPC(KNOT)                      4.671E-6
                  Object                         5.08E-8
                  Objective                     101354.5
                  Observations Read                   21
                  Observations Used                   21
                  Observations Missing                 0


                            Sum of        Mean               Approx
    Source          DF     Squares      Square    F Value    Pr > F

    Regression        4    2.9924E8    74810601    2144.72    <.0001
    Residual         17      101355      5962.0
    Uncorrected Total 21    2.9934E8

    Corrected Total  20    38462029


                                     Approx
    Parameter    Estimate    Std Error    Approximate 95% Confidence
                                                      Limits

    B0            -361.2      60.4110      -488.6       -233.7
    B1          61.1688       1.2144      58.6066      63.7310
    B2         -59.0143       1.6065     -62.4038     -55.6248
    KNOT        78.4859       1.1441      76.0722      80.8997


                    Approximate Correlation Matrix
                     B0            B1            B2            KNOT

    B0        1.0000000    -0.9146768     0.6914307     0.3965230
    B1       -0.9146768     1.0000000    -0.7559289    -0.5933197
    B2        0.6914307    -0.7559289     1.0000000     0.0353571
    KNOT      0.3965230    -0.5933197     0.0353571     1.0000000
```

Results in Output 7.12 provide the fitted model

$$\widehat{LEN29} = -361.2 + 61.17*AGE - 59.01*AGEPL$$

with the estimate of KNOT = 78.49. The value of KNOT can be incorporated into the equation as

$$\widehat{LEN29} = -361.2 + 61.17*AGE - 59.01*MAX(AGE - 78.49,0) \quad .$$

This provides the relationship

$$\widehat{LEN29} = -361.2 + 61.17*AGE$$

for AGE < 78.49 and

$$\overbrace{}$$
$$LEN29 = 4631.69 + 2.16*AGE$$

for AGE > 78.49, which is very close to the fitted linear spline model obtained from Output 6.7 with the knot fixed at 80. The spline model with estimated knot has SS(Error) = 101355 compared to 111756 for the model with the knot fixed at 80 as shown in Output 6.7.

In Chapter 6, "Special Applications of Linear Models," it is noted that splines with discontinuous derivatives are not universally acceptable. However, the quadratic spline with continuous derivatives cannot be justified for the fish growth data. Another alternative is a spline joining a straight line to an exponential curve, which is demonstrated next. (See "Example 45.1: Segmented Model" in "The NLIN Procedure," in the *SAS/STAT User's Guide* for an example of fitting a spline model joining a quadratic curve to a plateau.)

Recall that the exponential decay model fitted to the radioactive count data showed some lack of fit because the early decay appeared to be more linear than exponential. An alternative model has a straight line for TIME < t_0, and then an exponential decay curve for TIME > t_0. The equation for the curve for TIME > t_0 is

$$COUNT = B*\exp\left(C*TIME\right) + \varepsilon \quad .$$

The curve passes through the point (t_0, $B*\exp(C*t_0)$), and has slope $B*C*\exp(C*t_0)$ at this point. The straight line for TIME < t_0 must also pass through this point and have the same slope at this point. The point-slope form for the equation of a line (ignoring ε) gives

$$COUNT = B*\exp(C*t_0) + B*C*\exp(C*t_0)*(time - t_0)$$

as the equation for TIME < t_0. Equivalently,

$$COUNT = B*\exp(C*t_0)*(1 + C*(TIME - t_0)) \quad .$$

Fit this model with the following statements:

```
proc nlin data=decay;
   parms b=380 c=-.0028 t0=100;
      if time < t0 then do;
         model count = b*exp(c*t0)*(1 + c*(time - t0));
      end;
      else do;
   model count = b*exp(c*time);
      end;
   run;
```

It is logical to use the same starting values for B and C that were used when fitting the standard exponential decay curve (see Output 7.3). Results appear in Output 7.13.

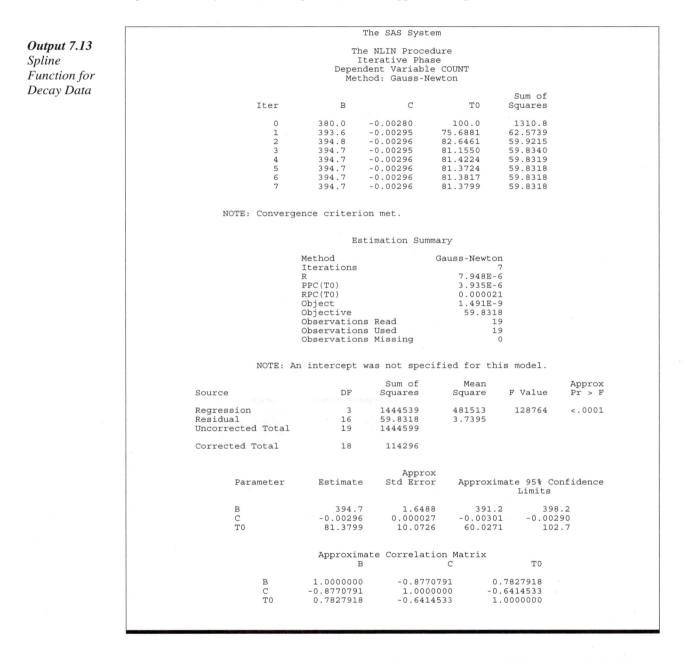

Output 7.13
Spline
Function for
Decay Data

```
                            The SAS System

                          The NLIN Procedure
                           Iterative Phase
                      Dependent Variable COUNT
                         Method: Gauss-Newton

                                                      Sum of
           Iter        B          C          T0       Squares

            0        380.0     -0.00280     100.0     1310.8
            1        393.6     -0.00295     75.6881    62.5739
            2        394.8     -0.00296     82.6461    59.9215
            3        394.7     -0.00295     81.1550    59.8340
            4        394.7     -0.00296     81.4224    59.8319
            5        394.7     -0.00296     81.3724    59.8318
            6        394.7     -0.00296     81.3817    59.8318
            7        394.7     -0.00296     81.3799    59.8318


     NOTE: Convergence criterion met.


                         Estimation Summary

                Method                  Gauss-Newton
                Iterations                        7
                R                          7.948E-6
                PPC(T0)                    3.935E-6
                RPC(T0)                    0.000021
                Object                     1.491E-9
                Objective                  59.8318
                Observations Read               19
                Observations Used               19
                Observations Missing             0


          NOTE: An intercept was not specified for this model.

                              Sum of       Mean                 Approx
     Source           DF      Squares      Square    F Value    Pr > F

     Regression        3      1444539      481513     128764     <.0001
     Residual         16      59.8318      3.7395
     Uncorrected Total 19     1444599

     Corrected Total   18     114296


                                     Approx
          Parameter   Estimate     Std Error   Approximate 95% Confidence
                                                         Limits

          B             394.7        1.6488       391.2       398.2
          C          -0.00296      0.000027     -0.00301    -0.00290
          T0          81.3799      10.0726       60.0271     102.7


                    Approximate Correlation Matrix
                           B            C            T0

              B       1.0000000   -0.8770791     0.7827918
              C      -0.8770791    1.0000000    -0.6414533
              T0      0.7827918   -0.6414533     1.0000000
```

Output 7.13 gives parameter estimates for the fitted model, along with SS(Residual)=59.83. This is a reduction in SS(Residual) from 143.06 for the standard exponential decay model shown in Output 7.3.

Add the following statements to obtain the residual plot and the plot of the fitted curve:

```
output out=plot p=pct r=rct;
proc plot data=plot vpercent=50 hpercent=50;
   plot rct*time;
   proc gplot;
   symbol1 v=star c=black;
   symbol2 v=point c=black i=spline l=1;
   plot count*time=1 pct*time=2 / overlay;
run;
```

The residual plot is shown in Output 7.14 and the plot of the observed values and the predicted line in Output 7.15.

Output 7.14
Residual
Plot for
Spline
Function

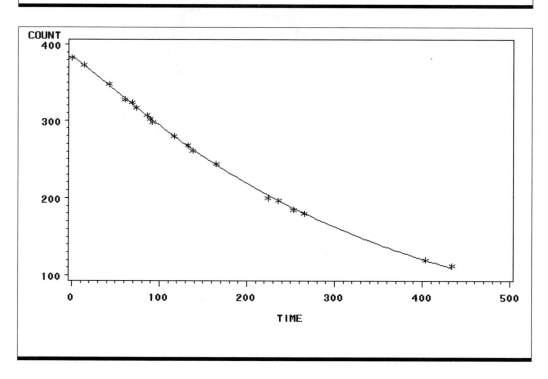

Output 7.15
Predicted Line
for Spline
Function

Output 7.14 displays no pattern in the residuals, and Output 7.15 shows the fitted curve passing so closely through observed data that the observed and predicted points are almost identical.

7.5 Additional Comments on the NLIN Procedure

PROC NLIN works quite well with the rather simple models used in this chapter. For more complicated models, especially models having many parameters, estimation is not always so straightforward. PROC NLIN provides a number of options and special features to assist in more difficult applications. Some of the ones most frequently used are outlined here.

7.5.1 Estimation Algorithms

PROC NLIN provides for several iterative algorithms for finding values of the parameters that minimize the residual sum of squares. Alternative algorithms may be implemented by using the appropriate METHOD= option in the PROC NLIN statement. Generally, the default Gauss-Newton method will work quite well, but the others may be used if there is a suspicion that the default has not provided a useful solution. All of these methods require the use of partial derivatives of the model function with respect to each of the parameters. However, unless the user provides the derivatives, they are calculated by the procedure.

For example, the derivatives for the decay model are

$$\frac{\partial \beta e^{\gamma t}}{\partial \beta} = e^{\gamma t}$$

and

$$\frac{\partial \beta e^{\gamma t}}{\partial_{\gamma}} = t\beta e^{\gamma t} \quad .$$

These derivatives are specified for PROC NLIN with DER statements:

```
proc nlin data = decay;
    parms b = 380 c = -.0026;
    model count = b*exp(c*time);
    der.b = exp(c*time);
    der.c = b*time*exp(c*time);
run;
```

The results are identical to those obtained previously and are therefore not shown here.

7.5.2 Other Methods

Difficulties with the estimation process increase much more rapidly with additional parameters in nonlinear regression than they do in linear regression models. For this reason, PROC NLIN provides a number of features, outlined below, to assist in providing estimates and other inferences that may be useful for more complicated models:

Initial values	are required for starting the iterative solution process and are not always easily obtained. As shown in the growth function model, the NLIN procedure allows for the implementation of a grid search method that computes the residual sum of squares for a grid of potential parameter values, and then starts iterations at that combination of values.
Convergence	is linked to choice of the initial values. Occasionally the output of the iterative process indicates that the convergence criterion has not been met. This means that a satisfactory least squares solution has not been found. Since this is usually the result of a poor choice of initial values, a grid search can be useful in determining better initial values. Specifying a different solution algorithm or allowing more iterations may also help. A number of options are available for modifying the iterative process.
Local minima	can be problematic if the solution found by the iterative procedure is local rather than the correct or global minimum residual sum of squares. Such an occurrence can result in an overly large residual sum of squares or unreasonable parameter estimates. One way to investigate this problem is to fit a polynomial model (for example, with PROC RSREG), which usually provides a reasonable approximation to many nonlinear functions, and compare the resulting residual mean square with that of PROC NLIN. If the PROC NLIN residual mean square is considerably larger, there may be a problem with the estimated nonlinear model. Again, different starting values or estimation methods should be tried if the estimates are in doubt.
Parameter estimates	can be restricted by using a BOUNDS statement. Implementation of BOUNDS statements may, however, invite finding local optima.
Programming statements	can be used between the PARMS and MODEL statements. Programming statements, such as those used in the DATA step, can simplify expressions and save computing time. For example, derivatives of exponential functions involve the same exponentials. Defining an exponential function once in a program step allows its repeated use without recomputation.

Specific instructions for using these various alternatives and options, together with other helpful hints, are found in the *SAS/STAT Users Guide*, which also contains several interesting examples including iteratively reweighted least squares and maximum-likelihood estimation.

7.6 Summary

This chapter has provided some relatively simple examples of how PROC NLIN can be used to estimate nonlinear regression models. Additional options for dealing with more complicated models have been briefly noted but not explicitly illustrated.

Fitting nonlinear models is not always an easy task, especially for models with four or more parameters. Special strategies may need to be considered.

The spline model was relatively easy to fit, even though it had four parameters, because three of the parameters were in linear form, and a closely related linear regression could be used to provide starting values. Therefore, if a nonlinear model contains a number of linear parameters, it may be possible to estimate the linear parameters by using reasonable values of the nonlinear parameters to get starting values.

As a rule, if you need to fit a rather complicated model and you are not very familiar with nonlinear models in general and PROC NLIN in particular, try some simple examples first.

Chapter 8

Using SAS/INSIGHT Software for Regression

8.1 Introduction

In this book you have learned how to use the SAS System to analyze data with various regression methods. You have done this by writing short programs using the SAS language that specify the desired analyses, options, and outputs. However, with the advent of windows based operating systems, many tasks that have traditionally been performed through such programs are now implemented by consulting a *menu* of possible tasks, and then *pointing* a *mouse* to the desired choice and *clicking* to implement the choice. For more complex tasks, a choice may bring up a new menu and the menu process continues.

This point-and-click approach to using computer programs has many advantages: You do not have to worry about the rules of various programming languages and the unforeseen consequences of misspelled words or typographical errors. It also tends to be more interactive; it is easier to quickly follow up an analysis with requests for additional information. On the other hand, the use of programs offers much more flexibility for specifying tasks and options. It would be very difficult to develop a user-friendly point-and-click approach that would encompass all the functions of the entire SAS System.

SAS/INSIGHT provides a point-and-click interactive tool for data exploration and analysis. The availability of SAS/INSIGHT as a SAS procedure provides you with the best of both worlds. For example, you can use the power of the DATA step to produce SAS data, invoke SAS/INSIGHT for some data exploration, and, if necessary, return to the SAS System to perform analyses not readily done with the point-and-click approach.

This chapter shows how you can use SAS/INSIGHT to perform some of the analyses presented in previous chapters. For more information, see the *SAS/INSIGHT User's Guide*.

8.2 Multiple Linear Regression: the BOQ Data

Assume that you have invoked the SAS System and have submitted the program to create the BOQ Data (Output 3.1).[1] You invoke SAS/INSIGHT as follows:

```
proc insight; run;
```

---

[1] Data may be entered directly into a PROC INSIGHT window.

Your screen will produce the window shown in Screen 8.1.

Screen 8.1
Initial
SAS/INSIGHT
Window

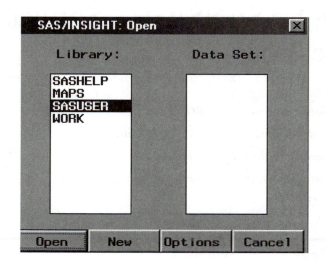

Initially, the right-hand side will be blank. The left-hand side shows available libraries that may contain data to be analyzed. Normally, the BOQ data set is in the temporary WORK library, so you select WORK, which brings up the data set BOQ. Select BOQ, and then Open, which results in Output 8.1. A comparison with Output 3.1 shows that the observations are identical.

Output 8.1
BOQ Data
for
SAS/INSIGHT
Software

	occup	checkin	hours	common	wings	cap	rooms	manh
1	2.00	4.00	4.0	1.26	1	6	6	180.23
2	3.00	1.58	40.0	1.25	1	5	5	182.61
3	5.30	1.67	42.5	7.79	3	25	25	199.92
4	7.00	2.37	168.0	1.00	1	7	8	284.55
5	16.50	8.25	168.0	1.12	2	19	19	267.38
6	16.60	23.78	40.0	1.00	1	13	13	164.38
7	25.89	3.00	40.0	0.00	3	36	36	999.09
8	31.92	40.08	168.0	5.52	6	47	47	931.84
9	39.63	50.86	40.0	27.37	10	77	77	944.21
10	44.42	159.75	168.0	0.60	18	48	48	1103.24
11	54.58	207.08	168.0	7.77	6	66	66	1387.82
12	56.63	373.42	168.0	6.03	4	36	37	1489.50
13	95.00	368.00	168.0	30.26	9	292	196	1845.89
14	96.67	206.67	168.0	17.86	14	120	120	1891.70
15	96.83	677.33	168.0	20.31	10	302	210	1880.84
16	97.33	255.08	168.0	19.00	6	165	130	2268.06
17	102.33	288.83	168.0	21.01	14	131	131	3036.63
18	110.24	410.00	168.0	20.05	12	115	115	2628.32
19	113.88	981.00	168.0	24.48	6	166	179	3559.92
20	134.32	145.82	168.0	25.99	12	192	192	2227.76
21	149.58	233.83	168.0	31.07	14	185	202	3115.29
22	188.74	937.00	168.0	45.44	26	237	237	4804.24
23	274.92	695.25	168.0	46.63	58	363	363	5539.98
24	384.50	1473.66	168.0	7.36	24	540	453	8266.77
25	811.08	714.33	168.0	22.76	17	242	242	3534.49

You are now ready to perform a regression analysis. Click Analyze from the menu bar, which produces the drop-down menu shown in Screen 8.2.

Screen 8.2
Analyze
Menu

Note that the menu includes several plot options; these will be presented later. Regression analyses are performed by clicking the Fit(YX) option, which produces the dialog box shown in Screen 8.3.

Screen 8.3
The Fit (YX)
Dialog Box

Initially the only entries in the screen consist of the list of variables in the data set BOQ on the left. From this list you will choose the variables for your model. Click MANH, the dependent variable, and then click Y. The selection of the set of independent variables can be done by any of the following: click them one at a time, hold down the CONTROL key while clicking the desired variables, hold down SHIFT while clicking OCCUP and ROOMS, or click and slide the cursor across the desired variables. Then click X and the independent variables are chosen. Screen 8.3 shows the results of these actions.

If you now click OK, a least squares regression with the usual outputs will be performed; you may obtain other estimation methods by clicking Method; these will be used in a later example. However, because you suspect multicollinearity, you can select additional output options by clicking Output, resulting in the dialog box shown in Screen 8.4.

Screen 8.4
Output
Options
Dialog
Box

Fit [Y X]

Tables:

☑ Model Equation
☐ X'X Matrix
☑ Summary of Fit
☑ Analysis of Variance/Deviance
☐ Type I / I(LR) Tests
☑ Type III / III(Wald) Tests
☐ Type III(LR) Tests
☑ Parameter Estimates
☐ 95% C.I. / C.I.(Wald) for Parameters
☐ 95% C.I.(LR) for Parameters
☐ Collinearity Diagnostics
☐ Estimated Cov Matrix
☐ Estimated Corr Matrix

Plots:

☑ Residual by Predicted
☐ Residual Normal QQ
☐ Partial Leverage

Surface Plots:

☑ Parametric
☐ Kernel (Normal GCV)
☐ Smoothing Spline (GCV)
☐ Parametric Profile

Parametric Curves

Output Variables

Nonparametric Curves (GCV)

OK   Cancel

The default output normally consists of the checked items but may be different because of the options chosen and saved from a prior analysis. Because you suspect multicollinearity, you can click Collinearity Diagnostics. Any chosen items may be deleted by clicking those items. If you have selected a residual-by-predicted plot, the predicted and residual values will be output to the data set, but if you want other plots, you will need to create the necessary variables for plotting. You can do this by clicking the Output Variables button, which produces the dialog box shown in Screen 8.5.

Screen 8.5
Output Variables Dialog Box

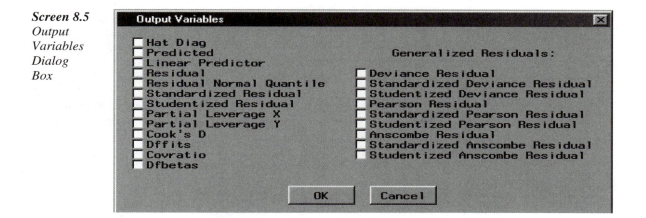

Output Variables

☐ Hat Diag
☐ Predicted   Generalized Residuals:
☐ Linear Predictor
☐ Residual   ☐ Deviance Residual
☐ Residual Normal Quantile   ☐ Standardized Deviance Residual
☐ Standardized Residual   ☐ Studentized Deviance Residual
☐ Studentized Residual   ☐ Pearson Residual
☐ Partial Leverage X   ☐ Standardized Pearson Residual
☐ Partial Leverage Y   ☐ Studentized Pearson Residual
☐ Cook's D   ☐ Anscombe Residual
☐ Dffits   ☐ Standardized Anscombe Residual
☐ Covratio   ☐ Studentized Anscombe Residual
☐ Dfbetas

[OK]   [Cancel]

Output 8.2
Result for Multiple Regression

MANH =   OCCUR CHECKIN HOURS COMMON WINGS CAP ROOMS
Response Distribution: Normal
Link Function: Identity

Model Equation

MANH =   134.968   -   1.2838   OCCUP +   1.8035   CHECKIN +   0.6692   HOURS
          - 21.4226   COMMON   5.6192   WINGS -   14.4803   CAP +   29.3248   ROOMS

Summary of Fit

Mean of Response	2109.3864	R-Square	0.9613
Root MSE	455.1673	Adj R-Sq	0.9453

Analysis of Variance

Source	DF	Sum of Squares	Mean Square	F Stat	Pr > F
Model	7	87387188.1	12483884.0	60.26	<.0001
Error	17	3522013.12	207177.242		
C Total	24	909092201.3			

Type III Tests

Source	DF	Sum of Squares	Mean Square	F Stat	Pr > F
OCCUP	1	527297.878	527297.878	2.55	0.1291
CHECKIN	1	2528612.68	2528612.68	12.21	0.0028
HOURS	1	27210.5928	27210.5928	0.13	0.7215
COMMON	1	918986.493	918986.493	4.44	0.0504
WINGS	1	30084.3815	30084.3815	0.15	0.7079
CAP	1	2439117.81	2439117.81	11.77	0.0032
ROOMS	1	4396333.14	4396333.14	21.22	0.0003

Parameter Estimates

Variable	DF	Estimate	Std Error	t Stat	Pr > |t|	Tolerance	Var Inflation
Intercept	1	134.9679	237.8143	0.57	0.5778	.	0
OCCUP	1	-1.2838	0.8047	-1.60	0.1291	0.4624	2.1628
CHECKIN	1	1.8035	0.5162	3.49	0.0028	0.2210	4.5240
HOURS	1	0.6692	1.8464	0.36	0.7215	0.7367	1.3573
COMMON	1	-21.4226	10.1716	-2.11	0.0504	0.4287	2.3326
WINGS	1	5.6192	14.7461	0.38	0.7079	0.2737	3.6532
CAP	1	-14.4803	4.2202	-3.43	0.0032	0.0269	37.1291
ROOMS	1	29.3248	6.3659	4.61	0.0003	0.0157	63.7081

Collinearity Diagnostics

						Variance Proportion				
Number	Eigenvalue	Condition Index	Intercept	OCCUP	CHECKIN	HOURS	COMMON	WINGS	CAP	ROOMS
1	6.4688	1.0000	0.0025	0.0049	0.0024	0.0021	0.0035	0.0028	0.0003	0.0002
2	0.6065	3.2657	0.0905	0.1039	0.0194	0.0392	0.0091	0.0008	0.0009	0.0004
3	0.3601	4.2385	0.0379	0.2348	0.0076	0.0215	0.1189	0.1198	9.296E-05	0.0002
4	0.2683	4.9103	0.0004	0.4683	0.1313	0.0012	0.0467	0.0460	0.0055	0.0009
5	0.1427	6.7335	0.0165	0.0095	0.0063	0.0042	0.6884	0.4413	0	2.855E-05
6	0.0810	8.9341	0.1222	1.004E-05	0.5258	0.1512	0.0185	0.0718	0.0506	0.0099
7	0.0680	9.7545	0.7293	5.271E-05	0.2201	0.7806	0.0089	0.0126	0.0084	0.0013
8	0.0046	37.4160	0.0007	0.1785	0.0870	2.103E-05	0.1059	0.3049	0.9342	0.9872

R-MANH vs P_MANH scatter plot with x-axis labeled 2000 4000 6000

Clicking OK provides the results of the regression procedure shown in Output 8.2. The results are the same as those found in Output 3.2; the multicollinearity diagnostics are somewhat different from those in Chapter 4, "Multicollinearity: Detection and Remedial Measures," because this analysis includes the "impossible" observation 25. You can print the results by selecting Print from the File menu.

A plot of DFFITS against the predicted values is done by a separate option. First close the Output window, which will return you to the data window. You will see that three new variables have been added, R_MANH, P_MANH, and F_MANH. These are the residual, predicted, and DFFITS values, respectively. Click the Analyze menu (Screen 8.2), and click Scatter Plot (Y X), which produces the dialog box shown in Screen 8.6.

Screen 8.6
Scatter Plot
Dialog Box

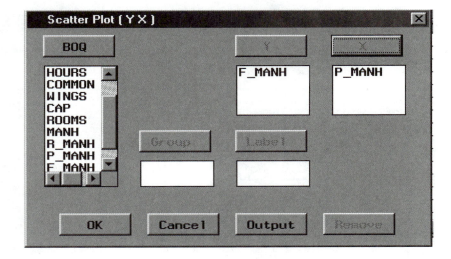

Select F_MANH as Y and P_MANH as X, and then click OK, which produces the graph shown in Output 8.3. The Output button provides some other options for the axes and labeling that are not described here.

Output 8.3
DFFITS
Plot

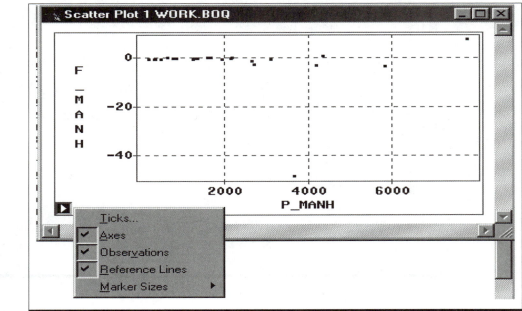

Initially, the plot does not have the reference lines. Click the arrow in the lower left corner, which produces the small menu shown in Output 8.3. Then click Reference Lines and the reference lines are produced. This plot is identical to the DFFITS plot in the lower-right-hand portion of Output 3.6.

The influential observation is obvious. You can identify that observation by clicking on it. The result, shown in Output 8.4, identifies it as observation 25. You can hold down the CONTROL key and click any number of observations. Clicking in a blank space will cancel the identification.

Output 8.4
Identifying
Influential
Observation

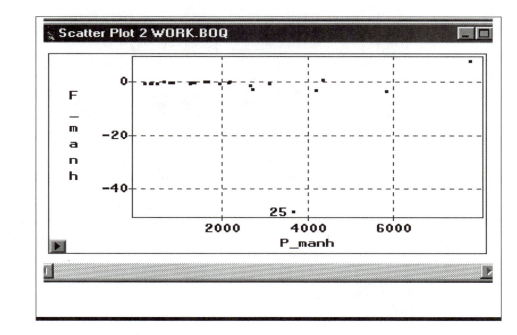

An alternate method for selecting observations is to click the Edit menu, then Observations, which produces another drop-down menu of choices, as shown in Screen 8.7.

Screen 8.7
Finding
Observations

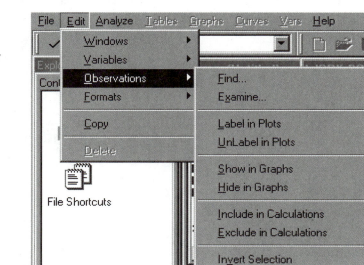

At this point you have several choices. If you want to identify observations, select Label in Plots, which produces the dialog box Show Observations, as in Screen 8.8.

Screen 8.8
Identifying
an
Observation

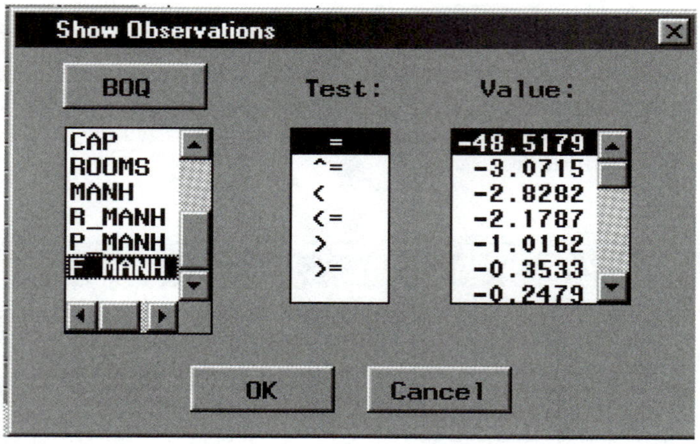

You can now select observations based on values of any of the variables in the data set. For example, scroll down the BOQ box until you find and click F_MANH, and the values of that statistic appear in the Value box. Scroll until the extreme value (−48.5179) appears, and click that value. Then select the equals sign (=) under TEST, and click OK. Note that you could select several values, such as all observations with values greater or less than a specified value. The resulting plot will be similar to the one shown in Output 8.4.

If you now want to do the analysis without that observation, return to the Edit: Observations menu. Continue with Exclude in Calculations and repeat the regression analysis.

Two interesting aspects of this data set are the existence of severe multicollinearity and the impossible observation 25. You can get a better idea of the interplay among these features by creating a scatter plot matrix, which is a matrix of pairwise scatter plots of a set of variables. To get such a plot, return to the BOQ data set dialog box. Select OCCUP, CAP, ROOMS, and MANH. Click the Analyze menu and then Scatter Plot (Y X). The result is shown in Output 8.5.

Output 8.5
Scatter Plot
Matrix

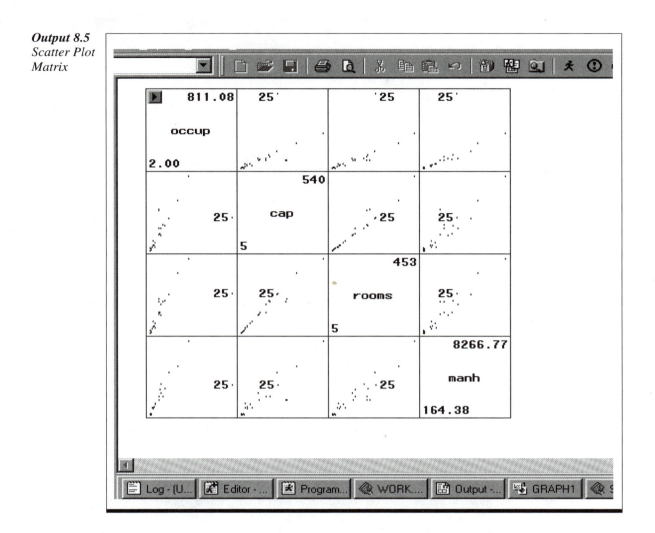

Output 8.5 shows all pairwise scatter plots of the selected variables. The numbers in the diagonal boxes define the extreme points of the axes for the variable in the corresponding row and column. For example, OCCUP has values from 2.00 to 811.08. Observation 25 is highlighted because this labeling option is still active. The very strong correlation between CAP and ROOMS is evident, as is the obviously suspicious value for OCCUP in observation 25.

8.3 A Polynomial Response Surface: the FISH Data

In Chapter 5, "Curve Fitting," the FISH data were used to illustrate the use of polynomial regression to produce curves and response surfaces. Having finished with the BOQ data, you can close the SAS/INSIGHT windows, return to SAS, and create the FISH data set. Invoke PROC INSIGHT, and the initial window will show two data sets, BOQ and FISH, in the WORK library. Open the FISH data and click Analyze: Fit (Y X), which results in Screen 8.9.

Screen 8.9
Fit Dialog
Box for FISH
Data

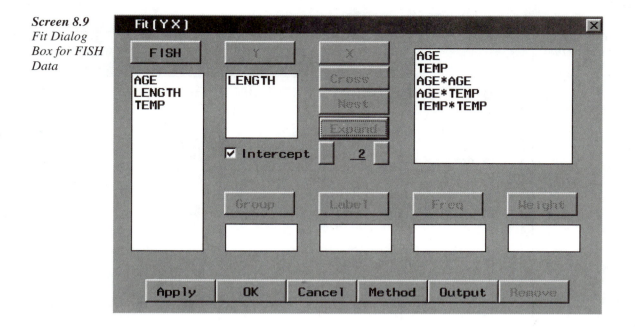

As before, click LENGTH and Y. Now select AGE and TEMP. Under X, you will see
CROSS, NEST, EXPAND, and the numeral 2. The 2 indicates that a polynomial of
degree 2 will be produced; higher orders are selected by clicking the plus (+) button to
the right. The Expand button creates products of all variables, including the polynomial
variables. Specifying polynomials of degree 2 and clicking the Expand button creates the
variables needed for the quadratic response surface as shown in the upper-right portion.[2]
Because you will want a response surface plot, click the Output button, which will
produce the dialog box shown in Screen 8.10.

Screen 8.10
Output
Options
Dialog Box

Fit [Y X]

Tables:
☑ Model Equation
☐ X'X Matrix
☑ Summary of Fit
☑ Analysis of Variance/Deviance
☐ Type I / I(LR) Tests
☑ Type III / III(Wald) Tests
☐ Type III(LR) Tests
☑ Parameter Estimates
☐ 95% C.I. / C.I.(Wald) for Parameters
☐ 95% C.I.(LR) for Parameters
☐ Collinearity Diagnostics
☐ Estimated Cov Matrix
☐ Estimated Corr Matrix

Plots:
☑ Residual by Predicted
☐ Residual Normal QQ
☐ Partial Leverage

Surface Plots:
☑ Parametric
☐ Kernel (Normal GCV)
☐ Smoothing Spline (GCV)
☐ Parametric Profile

Parametric Curves

Output Variables

Nonparametric Curves (GCV)

OK   Cancel

---

[2] Additional variables may be created by clicking the Edit menu and then clicking Variables, which produces a menu for adding
various transformations of the variables.

In response surface regressions, you are usually not interested in multicollinearity; hence the Collinearity Diagnostics button should not be checked. However, you will want the Parametric Surface Plot. Now click OK, return to the Fit (Y X) dialog box, and click OK. This produces the information shown in Output 8.6.

Output 8.6
Output for
Response
Surface
Regression

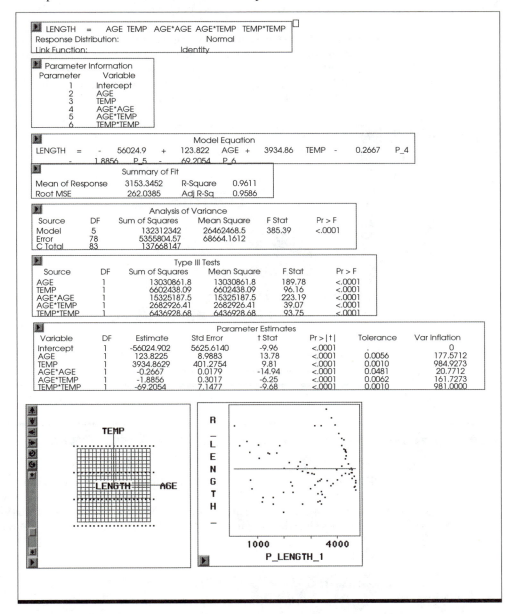

You can compare the results with those in Output 5.5 and see that they are identical. Of course, SAS/INSIGHT regards this as an ordinary multiple regression and therefore the special features of PROC RSREG are not provided. The residual plot also is the same as that shown in Output 5.7. However, the response surface plot seems to leave a lot to be desired. That is because the default perspective is from the top. You can rotate the plot by clicking on the various buttons at the top left of the plot. Specifically:

❑ the top two buttons rotate on the vertical axis

❑ the middle two buttons rotate on the horizontal axis

❑ the bottom two buttons rotate clockwise or counterclockwise.

Also the default for the axes places the origins at the means. You can change this by clicking on the small arrow at the bottom left, selecting axes, and clicking axes at minima. After playing with these controls for a few minutes, you may obtain the results in Output 8.7. The similarity to Output 5.8 is obvious except that the plot from SAS/INSIGHT also displays the actual response values.

Output 8.7
Rotated
Response
Surface Plot

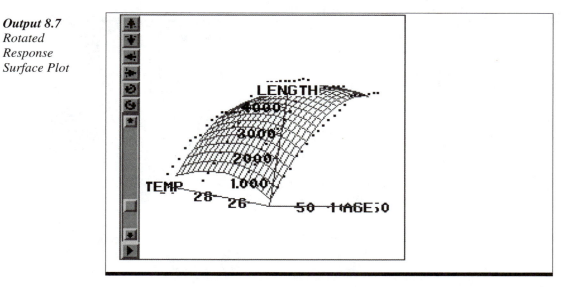

8.4 Logistic Regression: the DYSTRO Data

SAS/INSIGHT is not restricted to analyzing linear regression models by least squares methodology. You can, for example, perform a logistic regression, which was illustrated in Chapter 6, "Special Applications of Linear Models." Again, use the DATA step to create the DYSTRO data set, invoke PROC INSIGHT, and open the DYSTRO data set. Then click Analyze: Fit (YX), select CARRIER as Y and P1 through P4 as X, and click METHOD, resulting in the dialog box shown in Screen 8.11.

Screen 8.11
Fit Dialog Box
for Logistic
Regression

Click the Binomial response distribution and the Logit link function, which is the default used with PROC LOGISTIC. Click OK, return to the Fit (Y X) dialog box, and click the Output button. This produces the dialog box shown in Screen 8.12.

Screen 8.12
Output
Dialog Box
for Logistic
Regression

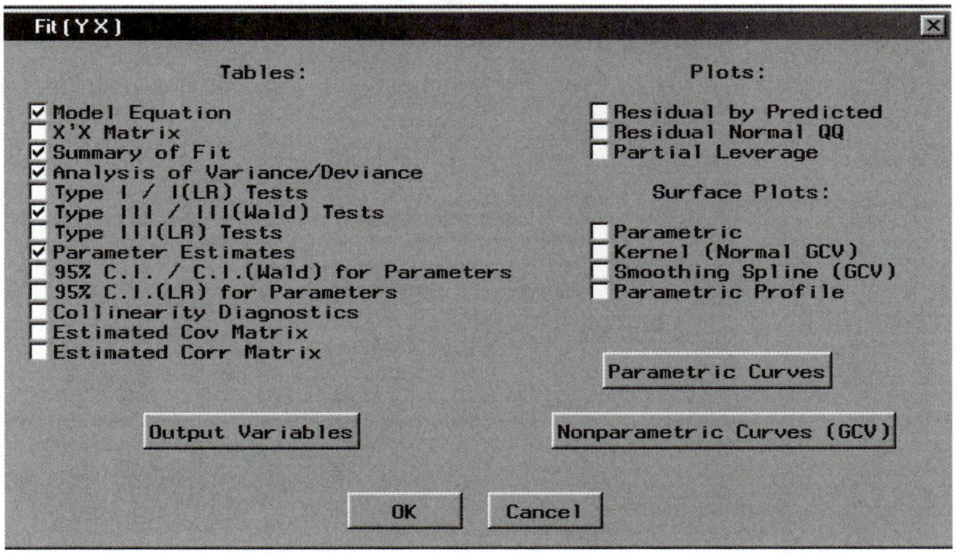

You will not need collinearity diagnostics or any plots. You may, however, want to output the predicted values for plotting. Click the Output Variables button and select Response Variable and Predicted. Click OK, which returns you to the Fit (Y X) dialog box, and click OK. This results in the information shown in Output 8.8.

Output 8.8
Logistic
Regression

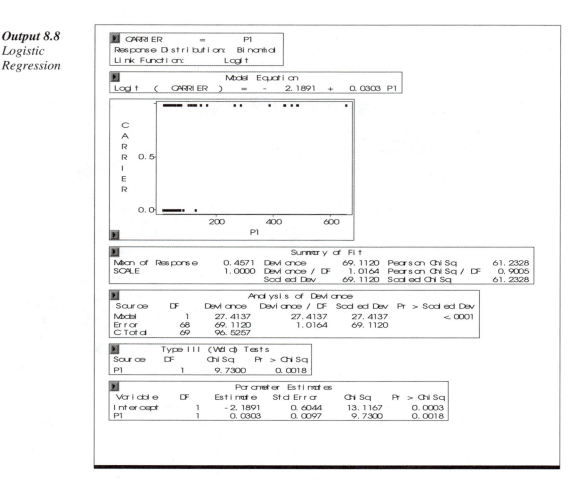

You can see that the results are the same as those that were obtained with PROC
LOGISTIC (Output 6.17), although some of the statistics are presented in a different
manner. For example, Deviance in the SAS/INSIGHT output is the –2 Log L statistic
from PROC LOGISTIC. To plot the predicted curve, return to the DYSTRO data set, and
click the Analyze menu, and then Scatter Plot (Y X). Screen 8.13 shows the results.

Screen 8.13
Scatter Plot
Dialog Box
for Logistic
Regression

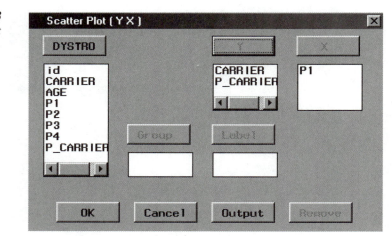

Choose CARRIER and P_CARRIER as Y, choose P1 as X, and click OK. The scatter plots are shown in Output 8.9.

Output 8.9
Logistic
Regression
Plots

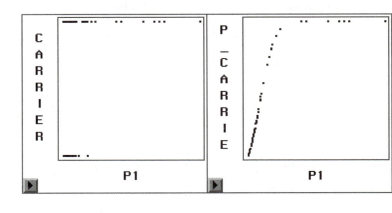

The response curve may be compared to Output 6.18, although you do not get the overlay option. You can, however, get a line plot for the predicted values by using the LINE PLOT option.

8.5 Nonparametric Smoothing: the Barbados Data

As well as fitting polynomial models, SAS/INSIGHT can also be used to perform nonparametric curve fitting. In Chapter 5, the Barbados exports data were used to illustrate how such analyses can be obtained with PROC LOESS. Assuming that the data set BARBEXP has been opened in SAS/ INSIGHT, click the Analyze menu, then Fit (Y X), which results in the dialog box shown in Screen 8.14.

Screen 8.14
Fit (YX)
Dialog Box
for
Barbados
Data

With SAS/INSIGHT, you will fit both a polynomial and nonparametric curve. First click EXPORT, then Y to define the response variable. Then click N (using YEAR for a polynomial regression may tend to cause round-off errors). The numeral under Expand controls the degree of polynomial to be used; the default is 2 (quadratic). Click the plus (+) button on the right to increase the power. To compare with the fifth-degree polynomial used in Chapter 5 (Output 5.16), click until a 5 appears. When you now click Expand, the screen will show the five powers of N as seen in the window.

To obtain a Loess fit, click Output (screen not shown), making sure that you have not chosen unnecessary outputs; then click Nonparametric Curves, which produces the dialog box shown in Screen 8.15.

Screen 8.15
Dialog Box
for LOESS Fit

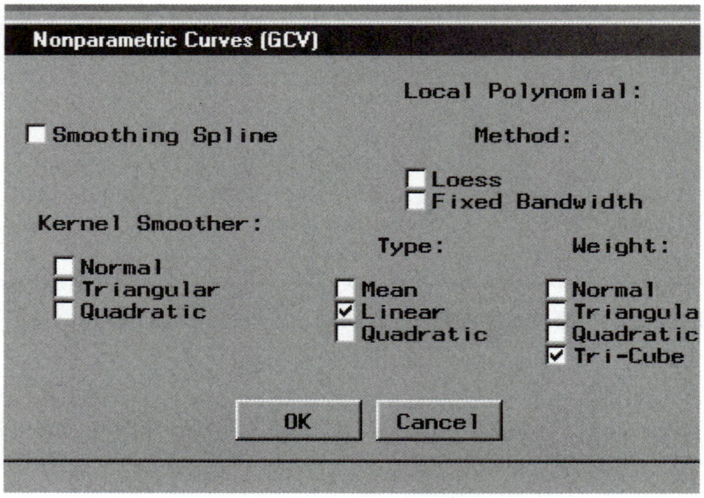

Click Loess for the Method, Linear for the Type, and Tri-Cube for the Weight. These options are phrased in a somewhat different manner from those for PROC LOESS. The value of the smoothing parameter to be used for PROC LOESS will be chosen later. Click OK on all dialog boxes, resulting in the information shown in Output 8.10.

Output 8.10
Polynomial and Loess Fit for Barbados Data

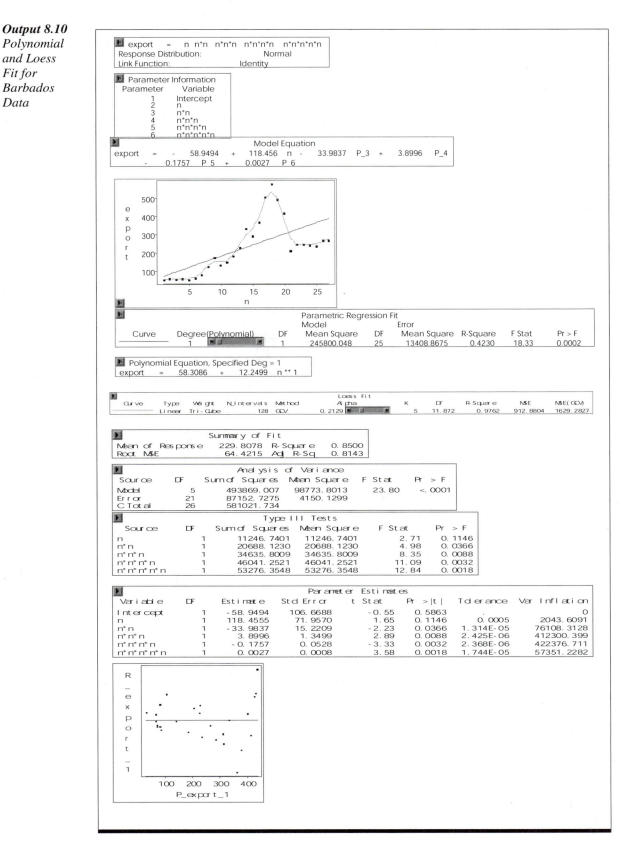

The first three and last four boxes provide information on the fifth-degree polynomial. You can verify that the results are the same as those that were obtained by PROC GLM (as reproduced in Output 5.17), except that SAS/INSIGHT does not provide the Type I sums of squares.

The fourth through sixth boxes provide for a comparison of polynomial and Loess fits. The plot gives the curves fitted by both models (on a computer monitor the curves are different colors) as specified in the next two boxes as follows:

❑   The straight line is provided by the linear (first-order) polynomial, whose statistics are given in the box immediately below the plot. The residual mean square is 13,409.

❑   The curved line is the Loess fit, whose statistics are shown in the sixth box. Remember that a Loess fit is a blending of weighted linear or quadratic curves in neighborhoods. In PROC LOESS the specification for the smoothing parameter is the fraction of observations in a neighborhood and is denoted here by Alpha. In SAS/INSIGHT the smoothing parameter is specified by the number of observations in a neighborhood that is denoted by K. The default used here is K = 5, which corresponds exactly to Alpha = 0.23. You can see that the curve compares favorably with the curve in Output 5.16 that uses the smoothing parameter 0.2. The residual mean square is 913, which implies a much better fit than that provided by the polynomial.

You can modify both curves interactively. Note the scroll bars beneath Degree (Polynomial) and Alpha in the output boxes. Clicking the right arrow (⇨) increases the corresponding parameter while clicking the left (⇦) decreases it. Alternately you can slide the scroll bar. As the parameters are changed, both the fit statistics and the plot will change. For example, a sixth-degree polynomial (a less smooth fit) and a Loess fit with K = 6 (a smoother fit) provide the results shown in Output 8.11.

Output 8.11
Changing
Smoothing
Parameters

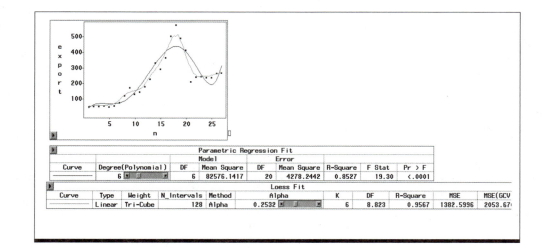

You can see that the curves have changed. The Loess fit is still better than that from the polynomial, but the difference between the two is smaller.

8.6 Summary

The purpose of this chapter was to show how SAS/INSIGHT can be used to implement some regression procedures in a point-and-click environment. In fact, most but not all of the methods illustrated in this book can be performed with SAS/INSIGHT. For example, principal component regression can be performed by using the Multivariate (Y X) option in the Analyze menu. On the other hand, variable selection is not available. Also SAS/INSIGHT does not provide for as much flexibility in presenting plots and graphs, but plots may be changed interactively.

Saving SAS/INSIGHT output and graphics is done by using Edit to select the items desired. Click File, select Save, then click Graphics File. A dialog box opens for selection of file format and name of file. If you simply want the numerical results, select Save, then click Tables. You can also use the PRINT SCREEN key and paste the image into a graphics processor.

References

Allen, D.M. (1970), "Mean Square Error of Prediction as a Criterion for Selecting Variables," *Technometrics*, 13, 469–475.

Belsley, D.A. (1984), "Demeaning Conditions Through Centering," followed by comments by R.D. Cook et al., *The American Statistician*, 38, 73–93.

Belsley, D.A., Kuh, E., and Welsch, R.E. (1980), *Regression Diagnostics*, New York: John Wiley & Sons, Inc.

Berk, K.N. (1977), "Tolerance and Condition in Regression Computations," *Journal of the American Statistical Association*, 72, 863–866.

Brocklebank, J.C. and Dickey, D.A. (1986), *SAS System for Forecasting Time Series*, Cary, NC: SAS Institute Inc.

Civil Aeronautics Board (August 1972), *Aircraft Operating Cost and Performance Report*, Washington, DC: U.S. Government Printing Office.

Dickens, J.W. and Mason, D.D. (1962), "A Peanut Sheller for Grading Samples: An Application in Statistical Design," *Transactions of the ASAE*, Volume 5, Number 1, 42–45.

Freund, R.J. and Minton, P.D. (1979), *Regression Methods: A Tool for Data Analysis*, New York: Marcel Dekker, Inc.

Freund, R.J. and Wilson, W.J. (1998), *Regression Analysis*, San Diego: Academic Press.

Fuller, W.A. (1978), *Introduction to Statistical Time Series*, New York: John Wiley & Sons, Inc.

Graybill, F. (1976), *Theory and Application of the Linear Model*, Boston: PWS and Kent Publishing Company, Inc.

Johnson, R.A. and Wichern, D.W. (1982), *Applied Multivariate Statistical Analysis*, Englewood Cliffs, NJ: Prentice Hall.

Kvålseth, T.O. (1985), "Cautionary Note about R^2," *The American Statistician*, 39, 279–286.

Littell, R.C., Freund, R.J., and Spector, P.C. (1991), *SAS System for Linear Models, Third Edition*, Cary, NC: SAS Institute Inc.

Mallows, C.P. (1973), "Some Comments on C(p)," *Technometrics*, 15, 661–675.

Montgomery, D.C. and Peck, E.A. (1982), *Introduction to Linear Regression Analysis*, New York: John Wiley & Sons, Inc.

Morrison, D.F. (1976), *Multivariate Statistical Methods, Second Edition*, New York: McGraw-Hill Book Co.

Myers, R.H. (1976), *Response Surface Methodology*, Blacksburg, VA: Virginia Polytechnic Institute and State University.

Myers, R.H. (1990), *Classical and Modern Regression with Applications, Second Edition*, Boston: PWS and Kent Publishing Company, Inc.

Neter, J., Wasserman, W., and Kutner, M.H. (1989), *Applied Linear Regression Models, Second Edition*, Homewood, IL: Richard D. Irwin Inc.

Rawlings, J.O., Pantula, S.G., and Dickey, D.A. (1998), *Applied Regression Analysis: A Research Tool, Second Edition,* New York: Springer Verlag.

Ryan, T.P., (1997), *Modern Regression Methods*, New York: John Wiley & Sons.

SAS Institute Inc. (1990), *SAS/ETS User's Guide, Version 6, First Edition,* Cary, NC: SAS Institute Inc.

SAS Institute Inc. (1990), *SAS/IML Software: Usage and Reference, Version 6, First Edition*, Cary, NC: SAS Institute Inc.

SAS Institute Inc. (1990), *SAS/STAT User's Guide, Version 6, Fourth Edition*, Volume 2, Cary, NC: SAS Institute Inc.

SAS Institute Inc. (1990), *SAS Language and Procedures: Usage, Version 6, First Edition*, Cary, NC: SAS Institute Inc.

SAS Institute Inc. (1990), *SAS Language: Reference, Version 6, First Edition*, Cary, NC: SAS Institute Inc.

SAS Institute Inc. (1990), *SAS Procedures Guide, Version 6, Third Edition*, Cary, NC: SAS Institute Inc.

SAS Institute Inc. (1999), *The Complete Guide to the SAS Output Delivery System, Version 8,* Cary, NC: SAS Institute Inc.

SAS Institute Inc. (1999), *SAS/INSIGHT User's Guide, Version 8,* Cary, NC: SAS Institute Inc.

Searle, S.R. (1971), *Linear Models*, New York: John Wiley & Sons, Inc.

Smith, P.L. (1979), "Splines as a Useful and Convenient Statistical Tool," *The American Statistician*, 33, 57–62.

Steel, R.G.B. and Torrie, J.H. (1980), *Principles and Procedures of Statistics, Second Edition*, New York: McGraw-Hill Book Co.

Index

236

Books and Tapes from SAS® Institute's Books by Users℠ program:

An Array of Challenges — Test Your SAS® Skills
by **Robert Virgile**

Annotate: Simply the Basics
by **Art Carpenter**

Applied Multivariate Statistics with SAS® Software, Second Edition
by **Ravindra Khattree**
and **Dayanand N. Naik**

Applied Statistics and the SAS® Programming Language, Fourth Edition
by **Ronald P. Cody**
and **Jeffrey K. Smith**

Beyond the Obvious with SAS® Screen Control Language
by **Don Stanley**

Carpenter's Complete Guide to the SAS® Macro Language
by **Art Carpenter**

The Cartoon Guide to Statistics
by **Larry Gonick**
and **Woollcott Smith**

Categorical Data Analysis Using the SAS® System
by **Maura E. Stokes, Charles S. Davis,**
and **Gary G. Koch**

Cody's Data Cleaning Techniques Using SAS® Software
by **Ron Cody**

Common Statistical Methods for Clinical Research with SAS® Examples
by **Glenn A. Walker**

Concepts and Case Studies in Data Management
by **William S. Calvert**
and **J. Meimei Ma**

Efficiency: Improving the Performance of Your SAS® Applications
by **Robert Virgile**

Essential Client/Server Survival Guide, Second Edition
by **Robert Orfali, Dan Harkey,**
and **Jeri Edwards**

Extending SAS® Survival Analysis Techniques for Medical Research
by **Alan Cantor**

A Handbook of Statistical Analyses Using SAS®
by **B.S. Everitt**
and **G. Der**

The How-To Book for SAS/GRAPH® Software
by **Thomas Miron**

In the Know ... SAS® Tips and Techniques From Around the Globe
by **Phil Mason**

Integrating Results through Meta-Analytic Review Using SAS® Software
by **Morgan C. Wang**
and **Brad J. Bushman**

Learning SAS® in the Computer Lab
by **Rebecca J. Elliott**

The Little SAS® Book: A Primer
by **Lora D. Delwiche**
and **Susan J. Slaughter**

The Little SAS® Book: A Primer, Second Edition
by **Lora D. Delwiche**
and **Susan J. Slaughter**
(updated to include Version 7 features)

Logistic Regression Using the SAS System: Theory and Application
by **Paul D. Allison**

Mastering the SAS® System, Second Edition
by **Jay A. Jaffe**

Multiple Comparisons and Multiple Tests Using the SAS® System
by **Peter H. Westfall, Randall D. Tobias,**
Dror Rom, Russell D. Wolfinger,
and **Yosef Hochberg**

The Next Step: Integrating the Software Life Cycle with SAS® Programming
by **Paul Gill**

Multivariate Data Reduction and Discrimination with SAS® Software
by **Ravindra Khattree**
and **Dayanand N. Naik**

Painless Windows 3.1: A Beginner's Handbook for SAS® Users
by **Jodie Gilmore**

Painless Windows: A Handbook for SAS® Users
by **Jodie Gilmore**
(for Windows NT and Windows 95)

Painless Windows: A Handbook for SAS® Users, Second Edition
by **Jodie Gilmore**
(updated to include Version 7 and Version 8 features)

PROC TABULATE by Example
by **Lauren E. Haworth**

*Professional SAS® Programmers Pocket Reference,
Second Edition*
by **Rick Aster**

Professional SAS® Programming Secrets, Second Edition
by **Rick Aster**
and **Rhena Seidman**

Professional SAS® User Interfaces
by **Rick Aster**

*Programming Techniques for Object-Based Statistical
Analysis with SAS® Software*
by **Tanya Kolosova**
and **Samuel Berestizhevsky**

Quick Results with SAS/GRAPH® Software
by **Arthur L. Carpenter**
and **Charles E. Shipp**

Quick Start to Data Analysis with SAS®
by **Frank C. Dilorio**
and **Kenneth A. Hardy**

*Reporting from the Field: SAS® Software Experts Present
Real-World Report-Writing
Applications*

SAS®Applications Programming: A Gentle Introduction
by **Frank C. Dilorio**

SAS® Foundations: From Installation to Operation
by **Rick Aster**

SAS® Macro Programming Made Easy
by **Michele M. Burlew**

SAS® Programming by Example
by **Ron Cody**
and **Ray Pass**

SAS® Programming for Researchers and Social Scientists
by **Paul E. Spector**

*SAS® Software Roadmaps: Your Guide to Discovering
the SAS® System*
by **Laurie Burch**
and **SherriJoyce King**

SAS® Software Solutions: Basic Data Processing
by **Thomas Miron**

*SAS® System for Elementary Statistical Analysis,
Second Edition*
by **Sandra D. Schlotzhauer**
and **Dr. Ramon C. Littell**

SAS® System for Forecasting Time Series, 1986 Edition
by **John C. Brocklebank**
and **David A. Dickey**

SAS® System for Linear Models, Third Edition
by **Ramon C. Littell, Rudolf J. Freund,**
and **Philip C. Spector**

SAS® System for Mixed Models
by **Ramon C. Littell, George A. Milliken, Walter W. Stroup,**
and **Russell D. Wolfinger**

SAS® System for Regression, Third Edition
by **Rudolf J. Freund**
and **Ramon C. Littell**

SAS® System for Statistical Graphics, First Edition
by **Michael Friendly**

SAS® Today! A Year of Terrific Tips
by **Helen Carey** and **Ginger Carey**

The SAS® Workbook and Solutions
(books in this set also sold separately)
by **Ron Cody**

*Selecting Statistical Techniques for Social Science Data:
A Guide for SAS® Users*
by **Frank M. Andrews, Laura Klem, Patrick M. O'Malley,
Willard L. Rodgers, Kathleen B. Welch,**
and **Terrence N. Davidson**

*Solutions for your GUI Applications Development Using
SAS/AF® FRAME Technology*
by **Don Stanley**

Statistical Quality Control Using the SAS® System
by **Dennis W. King, Ph.D.**

*A Step-by-Step Approach to Using the SAS® System
for Factor Analysis and Structural Equation Modeling*
by **Larry Hatcher**

*A Step-by-Step Approach to Using the SAS® System
for Univariate and Multivariate Statistics*
by **Larry Hatcher**
and **Edward Stepanski**

*Strategic Data Warehousing Principles Using
SAS® Software*
by **Peter R. Welbrock**

*Survival Analysis Using the SAS® System:
A Practical Guide*
by **Paul D. Allison**

*Table-Driven Strategies for Rapid SAS® Applications
Development*
by **Tanya Kolosova**
and **Samuel Berestizhevsky**

Tuning SAS® Applications in the MVS Environment
by **Michael A. Raithel**

*Univariate and Multivariate General Linear Models:
Theory and Applications Using SAS® Software*
by **Neil H. Timm**
and **Tammy A. Mieczkowski**

Visualizing Categorical Data
by **Michael Friendly**

Working with the SAS® System
by **Erik W. Tilanus**

Your Guide to Survey Research Using the SAS® System
by **Archer Gravely**

JMP® Books

Basic Business Statistics: A Casebook
by **Dean P. Foster, Robert A. Stine,**
and **Richard P. Waterman**

Business Analysis Using Regression: A Casebook
by **Dean P. Foster, Robert A. Stine,**
and **Richard P. Waterman**

JMP® Start Statistics, Version 3
by **John Sall** *and* **Ann Lehman**

WILEY SERIES IN PROBABILITY AND STATISTICS

ESTABLISHED BY WALTER A. SHEWHART AND SAMUEL S. WILKS

Editors
Noel A. C. Cressie, Nicholas I. Fisher, Iain M. Johnstone, J. B. Kadane,
David W. Scott, Bernard W. Silverman, Adrian F. M. Smith,
Jozef L. Teugels; Vic Barnett, Emeritus, Ralph A. Bradley, Emeritus,   J. Stuart Hunter, Emeritus, David G.
Kendall, Emeritus

Probability and Statistics Section

*ANDERSON · The Statistical Analysis of Time Series
 ARNOLD, BALAKRISHNAN, and NAGARAJA · A First Course in Order Statistics
 ARNOLD, BALAKRISHNAN, and NAGARAJA · Records
 BACCELLI, COHEN, OLSDER, and QUADRAT · Synchronization and Linearity: An Algebra for Discrete Event Systems
 BARNETT · Comparative Statistical Inference, *Third Edition*
 BASILEVSKY · Statistical Factor Analysis and Related Methods: Theory and Applications
 BERNARDO and SMITH · Bayesian Statistical Concepts and Theory
 BILLINGSLEY · Convergence of Probability Measures, *Second Edition*
 BOROVKOV · Asymptotic Methods in Queuing Theory
 BOROVKOV · Ergodicity and Stability of Stochastic Processes
 BRANDT, FRANKEN, and LISEK · Stationary Stochastic Models
 CAINES · Linear Stochastic Systems
 CAIROLI and DALANG · Sequential Stochastic Optimization
 CONSTANTINE · Combinatorial Theory and Statistical Design
 COOK · Regression Graphics
 COVER and THOMAS · Elements of Information Theory
 CSÖRGŐ and HORVÁTH · Weighted Approximations in Probability Statistics
 CSÖRGŐ and HORVÁTH · Limit Theorems in Change Point Analysis
*DANIEL · Fitting Equations to Data: Computer Analysis of Multifactor Data, *Second Edition*
 DETTE and STUDDEN · The Theory of Canonical Moments with Applications in Statistics, Probability, and Analysis
 DEY and MUKERJEE · Fractional Factorial Plans
*DOOB · Stochastic Processes
 DRYDEN and MARDIA · Statistical Shape Analysis
 DUPUIS and ELLIS · A Weak Convergence Approach to the Theory of Large Deviations
 ETHIER and KURTZ · Markov Processes: Characterization and Convergence
 FELLER · An Introduction to Probability Theory and Its Applications, Volume I, *Third Edition,* Revised;
    Volume II, *Second Edition*
 FULLER · Introduction to Statistical Time Series, *Second Edition*
 FULLER · Measurement Error Models
 GHOSH, MUKHOPADHYAY, and SEN · Sequential Estimation
 GIFI · Nonlinear Multivariate Analysis
 GUTTORP · Statistical Inference for Branching Processes
 HALL · Introduction to the Theory of Coverage Processes
 HAMPEL · Robust Statistics: The Approach Based on Influence Functions
 HANNAN and DEISTLER · The Statistical Theory of Linear Systems
 HUBER · Robust Statistics
 HUSKOVA, BERAN, and DUPAC · Collected Works of Jaroslav Hajek—with Commentary
 IMAN and CONOVER · A Modern Approach to Statistics
 JUREK and MASON · Operator-Limit Distributions in Probability Theory
 KASS and VOS · Geometrical Foundations of Asymptotic Inference
 KAUFMAN and ROUSSEEUW · Finding Groups in Data: An Introduction to Cluster Analysis
 KELLY · Probability, Statistics, and Optimization
 KENDALL, BARDEN, CARNE, and LE · Shape and Shape Theory
 LINDVALL · Lectures on the Coupling Method
 McFADDEN · Management of Data in Clinical Trials
 MANTON, WOODBURY, and TOLLEY · Statistical Applications Using Fuzzy Sets
 MORGENTHALER and TUKEY · Configural Polysampling: A Route to Practical Robustness
 MUIRHEAD · Aspects of Multivariate Statistical Theory
 OLIVER and SMITH · Influence Diagrams, Belief Nets and Decision Analysis
*PARZEN · Modern Probability Theory and Its Applications
 PEÑA, TIAO, and TSAY · A Course in Time Series Analysis
 PRESS · Bayesian Statistics: Principles, Models, and Applications
 PUKELSHEIM · Optimal Experimental Design

*Now available in a lower priced paperback edition in the Wiley Classics Library.

*Now available in a lower priced paperback edition in the Wiley Classics Library.

*Now available in a lower priced paperback edition in the Wiley Classics Library.

Applied Probability and Statistics (Continued)

MALLER and ZHOU · Survival Analysis with Long Term Survivors

MANN, SCHAFER, and SINGPURWALLA · Methods for Statistical Analysis of Reliability and Life Data

McLACHLAN · Discriminant Analysis and Statistical Pattern Recognition

McLACHLAN and KRISHNAN · The EM Algorithm and Extensions

McLACHLAN and PEEL · Finite Mixture Models

McNEIL · Epidemiological Research Methods

MEEKER and ESCOBAR · Statistical Methods for Reliability Data

*MILLER · Survival Analysis, *Second Edition*

MONTGOMERY and PECK · Introduction to Linear Regression Analysis, *Second Edition*

MYERS and MONTGOMERY · Response Surface Methodology: Process and Product in Optimization Using Designed Experiments

NELSON · Accelerated Testing, Statistical Models, Test Plans, and Data Analyses

NELSON · Applied Life Data Analysis

OCHI · Applied Probability and Stochastic Processes in Engineering and Physical Sciences

OKABE, BOOTS, and SUGIHARA · Spatial Tesselations: Concepts and Applications of Voronoi Diagrams

PANKRATZ · Forecasting with Dynamic Regression Models

PANKRATZ · Forecasting with Univariate Box-Jenkins Models: Concepts and Cases

PIANTADOSI · Clinical Trials: A Methodologic Perspective

PORT · Theoretical Probability for Applications

PUTERMAN · Markov Decision Processes: Discrete Stochastic Dynamic Programming

RACHEV · Probability Metrics and the Stability of Stochastic Models

RÉNYI · A Diary on Information Theory

RIPLEY · Spatial Statistics

RIPLEY · Stochastic Simulation

ROLSKI, SCHMIDLI, SCHMIDT, and TEUGELS · Stochastic Processes for Insurance and Finance

ROUSSEEUW and LEROY · Robust Regression and Outlier Detection

RUBIN · Multiple Imputation for Nonresponse in Surveys

RUBINSTEIN · Simulation and the Monte Carlo Method

RUBINSTEIN and MELAMED · Modern Simulation and Modeling

RYAN · Statistical Methods for Quality Improvement, *Second Edition*

SCHIMEK · Smoothing and Regression: Approaches, Computation, and Application

SCHUSS · Theory and Applications of Stochastic Differential Equations

SCOTT · Multivariate Density Estimation: Theory, Practice, and Visualization

*SEARLE · Linear Models

SEARLE · Linear Models for Unbalanced Data

SEARLE, CASELLA, and McCULLOCH · Variance Components

SENNOTT · Stochastic Dynamic Programming and the Control of Queueing Systems

STOYAN, KENDALL, and MECKE · Stochastic Geometry and Its Applications, *Second Edition*

STOYAN and STOYAN · Fractals, Random Shapes and Point Fields: Methods of Geometrical Statistics

THOMPSON · Empirical Model Building

THOMPSON · Sampling

THOMPSON · Simulation: A Modeler's Approach

TIJMS · Stochastic Modeling and Analysis: A Computational Approach

TIJMS · Stochastic Models: An Algorithmic Approach

TITTERINGTON, SMITH, and MAKOV · Statistical Analysis of Finite Mixture Distributions

UPTON and FINGLETON · Spatial Data Analysis by Example, Volume 1: Point Pattern and Quantitative Data

UPTON and FINGLETON · Spatial Data Analysis by Example, Volume II: Categorical and Directional Data

VAN RIJCKEVORSEL and DE LEEUW · Component and Correspondence Analysis

VIDAKOVIC · Statistical Modeling by Wavelets

WEISBERG · Applied Linear Regression, *Second Edition*

WESTFALL and YOUNG · Resampling-Based Multiple Testing: Examples and Methods for p-Value Adjustment

WHITTLE · Systems in Stochastic Equilibrium

WOODING · Planning Pharmaceutical Clinical Trials: Basic Statistical Principles

WOOLSON · Statistical Methods for the Analysis of Biomedical Data

*ZELLNER · An Introduction to Bayesian Inference in Econometrics

Texts and References Section

AGRESTI · An Introduction to Categorical Data Analysis

ANDERSON · An Introduction to Multivariate Statistical Analysis, *Second Edition*

ANDERSON and LOYNES · The Teaching of Practical Statistics

ARMITAGE and COLTON · Encyclopedia of Biostatistics: Volumes 1 to 6 with Index

BARTOSZYNSKI and NIEWIADOMSKA-BUGAJ · Probability and Statistical Inference

BENDAT and PIERSOL · Random Data: Analysis and Measurement Procedures, *Third Edition*

*Now available in a lower priced paperback edition in the Wiley Classics Library.

Texts and References (Continued)

BERRY, CHALONER, and GEWEKE · Bayesian Analysis in Statistics and Econometrics: Essays in Honor of Arnold Zellner

BHATTACHARYA and JOHNSON · Statistical Concepts and Methods

BILLINGSLEY · Probability and Measure, *Second Edition*

BOX · R. A. Fisher, the Life of a Scientist

BOX, HUNTER, and HUNTER · Statistics for Experimenters: An Introduction to Design, Data Analysis, and Model Building

BOX and LUCEÑO · Statistical Control by Monitoring and Feedback Adjustment

BROWN and HOLLANDER · Statistics: A Biomedical Introduction

CHATTERJEE and PRICE · Regression Analysis by Example, *Third Edition*

COOK and WEISBERG · Applied Regression Including Computing and Graphics

COOK and WEISBERG · An Introduction to Regression Graphics

COX · A Handbook of Introductory Statistical Methods

DILLON and GOLDSTEIN · Multivariate Analysis: Methods and Applications

*DODGE and ROMIG · Sampling Inspection Tables, *Second Edition*

DRAPER and SMITH · Applied Regression Analysis, *Third Edition*

DUDEWICZ and MISHRA · Modern Mathematical Statistics

DUNN and CLARK · Basic Statistics: A Primer for the Biomedical Sciences, *Third Edition*

EVANS, HASTINGS, and PEACOCK · Statistical Distributions, *Third Edition*

FISHER and VAN BELLE · Biostatistics: A Methodology for the Health Sciences

FREEMAN and SMITH · Aspects of Uncertainty: A Tribute to D. V. Lindley

GROSS and HARRIS · Fundamentals of Queueing Theory, *Third Edition*

HALD · A History of Probability and Statistics and their Applications Before 1750

HALD · A History of Mathematical Statistics from 1750 to 1930

HELLER · MACSYMA for Statisticians

HOEL · Introduction to Mathematical Statistics, *Fifth Edition*

HOLLANDER and WOLFE · Nonparametric Statistical Methods, *Second Edition*

HOSMER and LEMESHOW · Applied Logistic Regression, *Second Edition*

HOSMER and LEMESHOW · Applied Survival Analysis: Regression Modeling of Time to Event Data

JOHNSON and BALAKRISHNAN · Advances in the Theory and Practice of Statistics: A Volume in Honor of Samuel Kotz

JOHNSON and KOTZ (editors) · Leading Personalities in Statistical Sciences: From the Seventeenth Century to the Present

JUDGE, GRIFFITHS, HILL, LÜTKEPOHL, and LEE · The Theory and Practice of Econometrics, *Second Edition*

KHURI · Advanced Calculus with Applications in Statistics

KOTZ and JOHNSON (editors) · Encyclopedia of Statistical Sciences: Volumes 1 to 9 with Index

KOTZ and JOHNSON (editors) · Encyclopedia of Statistical Sciences: Supplement Volume

KOTZ, REED, and BANKS (editors) · Encyclopedia of Statistical Sciences: Update Volume 1

KOTZ, REED, and BANKS (editors) · Encyclopedia of Statistical Sciences: Update Volume 2

LAMPERTI · Probability: A Survey of the Mathematical Theory, *Second Edition*

LARSON · Introduction to Probability Theory and Statistical Inference, *Third Edition*

LE · Applied Categorical Data Analysis

LE · Applied Survival Analysis

MALLOWS · Design, Data, and Analysis by Some Friends of Cuthbert Daniel

MARDIA · The Art of Statistical Science: A Tribute to G. S. Watson

MASON, GUNST, and HESS · Statistical Design and Analysis of Experiments with Applications to Engineering and Science

McCULLOCH and SEARLE · Generalized, Linear, and Mixed Models

MURRAY · X-STAT 2.0 Statistical Experimentation, Design Data Analysis, and Nonlinear Optimization

PURI, VILAPLANA, and WERTZ · New Perspectives in Theoretical and Applied Statistics

RENCHER · Linear Models in Statistics

RENCHER · Methods of Multivariate Analysis

RENCHER · Multivariate Statistical Inference with Applications

ROSS · Introduction to Probability and Statistics for Engineers and Scientists

ROHATGI · An Introduction to Probability Theory and Mathematical Statistics

ROHATGI and SALEH · An Introduction to Probability and Statistics, *Second Edition*

RYAN · Modern Regression Methods

SCHOTT · Matrix Analysis for Statistics

SEARLE · Matrix Algebra Useful for Statistics

STYAN · The Collected Papers of T. W. Anderson: 1943–1985

TIAO, BISGAARD, HILL, PEÑA, and STIGLER (editors) · Box on Quality and Discovery: with Design, Control, and Robustness

TIERNEY · LISP-STAT: An Object-Oriented Environment for Statistical Computing and Dynamic Graphics

WONNACOTT and WONNACOTT · Econometrics, *Second Edition*

WU and HAMADA · Experiments: Planning, Analysis, and Parameter Design Optimization

*Now available in a lower priced paperback edition in the Wiley Classics Library.

WILEY SERIES IN PROBABILITY AND STATISTICS
ESTABLISHED BY WALTER A. SHEWHART AND SAMUEL S. WILKS

Editors
Robert M. Groves, Graham Kalton, J. N. K. Rao, Norbert Schwarz, Christopher Skinner

Survey Methodology Section

*Now available in a lower priced paperback edition in the Wiley Classics Library.